NIST NCSTAR 1-3B

Federal Building and Fire Safety Investigation of the World Trade Center Disaster

Steel Inventory and Identification

Stephen W. Banovic
Materials Science and Engineering Laboratory
National Institute of Standards and Technology

September 2005

U.S. Department of Commerce
Carlos M. Gutierrez, Secretary

Technology Administration
Michelle O'Neill, Acting Under Secretary for Technology

National Institute of Standards and Technology
William Jeffrey, Director

Disclaimer No. 1

Certain commercial entities, equipment, products, or materials are identified in this document in order to describe a procedure or concept adequately or to trace the history of the procedures and practices used. Such identification is not intended to imply recommendation, endorsement, or implication that the entities, products, materials, or equipment are necessarily the best available for the purpose. Nor does such identification imply a finding of fault or negligence by the National Institute of Standards and Technology.

Disclaimer No. 2

The policy of NIST is to use the International System of Units (metric units) in all publications. In this document, however, units are presented in metric units or the inch-pound system, whichever is prevalent in the discipline.

Disclaimer No. 3

Pursuant to section 7 of the National Construction Safety Team Act, the NIST Director has determined that certain evidence received by NIST in the course of this Investigation is "voluntarily provided safety-related information" that is "not directly related to the building failure being investigated" and that "disclosure of that information would inhibit the voluntary provision of that type of information" (15 USC 7306c).

In addition, a substantial portion of the evidence collected by NIST in the course of the Investigation has been provided to NIST under nondisclosure agreements.

Disclaimer No. 4

NIST takes no position as to whether the design or construction of a WTC building was compliant with any code since, due to the destruction of the WTC buildings, NIST could not verify the actual (or as-built) construction, the properties and condition of the materials used, or changes to the original construction made over the life of the buildings. In addition, NIST could not verify the interpretations of codes used by applicable authorities in determining compliance when implementing building codes. Where an Investigation report states whether a system was designed or installed as required by a code *provision*, NIST has documentary or anecdotal evidence indicating whether the requirement was met, or NIST has independently conducted tests or analyses indicating whether the requirement was met.

Use in Legal Proceedings

No part of any report resulting from a NIST investigation into a structural failure or from an investigation under the National Construction Safety Team Act may be used in any suit or action for damages arising out of any matter mentioned in such report (15 USC 281a; as amended by P.L. 107-231).

ABSTRACT

As a result of the recovery efforts of the Structural Engineers Association of New York, Federal Emergency Management Agency/American Society of Civil Engineers, and the National Institute of Standards and Technology (NIST), NIST possesses 236 structural steel elements from the World Trade Center (WTC) buildings. These samples include full exterior column panels, core columns, portions of the floor truss members, channels used to attach the floor trusses to the interior columns, and other smaller structural components (e.g., bolts, diagonal bracing straps, aluminum façade). Many significant pieces were recovered from the impact and fire-affected floors. Additionally, the recovered structural elements have yielded sufficient representative samples, with respect to the determination of the quality and mechanical properties of the steel, for all 12 grades of exterior panel material, 2 grades of the core column material (representing 99 percent, by total number, of the columns), and both grades for the floor truss material. The lack of WTC 7 steel precludes tests on actual material from the structure; however, WTC 7 was constructed of three grades of conventional steel (36 ksi, 42 ksi, and 50 ksi), and literature values may be used to estimate properties.

Keywords: Identification, inventory, recovered, steel, structural elements, World Trade Center.

TABLE OF CONTENTS

Abstract iii
List of Figures vii
List of Tables ix
List of Acronyms and Abbreviations xi
Preface xiii
Acknowledgments xxiii
Executive Summary xxv

Chapter 1
Introduction 1
    1.1 Purpose of Report 1
    1.2 Scope of Report 1

Chapter 2
Background Information Related to Recovery of WTC Structural Steel 3

Chapter 3
Structural Elements Recovered from the WTC Buildings 5
    3.1 Present Location and Labeling of Structural Steel Elements 5
    3.2 Identification of WTC Structural Steel Elements 7

Chapter 4
Structural Steel Elements of Special Importance 47
    4.1 Samples Located in or Around the Floors Impacted by the Airplane 47
    4.2 Samples Representing the Various Types of Steel Specified in the Design Drawings 49

Chapter 5
Summary 55

Chapter 6
References 57

Appendix A
Data on Recovered WTC Steel

LIST OF FIGURES

Figure P–1. The eight projects in the federal building and fire safety investigation of the WTC disaster. xv

Figure 3–1. Characteristic "overall" view of the samples taken for each piece received. Sample shown here is C-14. 6

Figure 3–2. Location of the exterior panels recovered from the top third of WTC 1 and WTC 2. 9

Figure 3–3. Example of stampings on the interior base of the middle column for each panel. 11

Figure 3–4. Example of stampings placed on one end of a core column. 12

Figure 3–5. (a) Example of stamping placed on flange indicating the column type (120), and (b) schematic indicating the various plates corresponding to Table 3–5. 13

Figure 3–6. (a) Characteristic stenciling found on the lower portions of the exterior column panels for sample C-14. (b) Characteristic stenciling found on an interior core column for sample B-6152. 15

Figure 3–7. Schematic showing derrick divisions that hoisted the specific columns for (a) WTC 1 and (b) WTC 2. The "x" signifies the information that was not readable. 16

Figure 3–8. Schematic showing the sample M-10 as two separate exterior column panels, M-10a and M-10b. 25

Figure 3–9. Schematics displaying the various types of exterior column panels. 27

Figure 3–10. Exterior column panel maps indicating the portion of the specific exterior column panel section recovered from WTC 1. 28

Figure 3–11. Exterior column panel maps indicating the portion of the specific exterior column panel section recovered from WTC 2. 35

Figure 3–12. Core columns recovered from WTC 1. a) B-1011 (508A: 51–54), lower 2 ft to 3 ft of built-up box column, b) B-6152-1 (803A: 15–18), lower 3 ft of built-up box column. 39

Figure 3–13. Core columns recovered from WTC 2. a) C-88a (801B: 80–83), lower 16 ft of built-up box column and C-88b (801B: 77–80), upper 8 ft of built-up box column. 43

Figure 3–14. Structural element composed of three wide flange sections bolted together. The component was found to be from the framed floor area outside of the core on the 107th floor of WTC 1 (sample was C-26). 45

Figure 4–1. Interpreted column damage, from photographic evidence, to WTC 1, with overlay of samples in NIST's possession. Samples shown represent recovered portions. Core columns 603 and 605 are in the second row from the north face of WTC 1. 48

Figure 4–2. Interpreted column damage, from photographic evidence, to WTC 2, with overlay of samples in NIST's possession. 49

LIST OF TABLES

Table P–1.	Federal building and fire safety investigation of the WTC disaster.	xiv
Table P–2.	Public meetings and briefings of the WTC Investigation.	xvii
Table 3–1.	Identified exterior column panel pieces from WTC 1 and WTC 2.	8
Table 3–2.	Partially identified exterior column panel from WTC 1 or WTC 2.	9
Table 3–3.	Identified pieces of core column material from WTC 1 and WTC 2.	10
Table 3–4.	Other built-up box columns and wide flange sections from WTC 1 and WTC 2 with ambiguous stampings and/or markings.	10
Table 3–5.	Examples of column types with corresponding plate gauges.	14
Table 3–6.	Specified and observed minimum yield strengths for positively identified exterior column panels.	18
Table 3–7.	Specified and observed column types for positively identified exterior column panels.	19
Table 3–8.	Specified minimum yield strengths from WTC 1 and WTC 2, along with the observed stampings, used to positively identify some exterior column panels.	20
Table 3–9.	Specified column types of exterior panels from WTC 1 and WTC 2, along with the observed stampings, used to positively identify some exterior column panels.	21
Table 3–10.	Information used to determine the identification of exterior panel ASCE-2.	22
Table 3–11.	Information used to determine the identification of exterior panel M-2.	24
Table 4–1.	Listing of recovered exterior column panels with specified minimum yield strengths and thickness for columns and spandrels.	51
Table 4–2.	Strength/gauge combinations of perimeter columns recovered by NIST.	52
Table 4–3.	Strength/gauge combinations of spandrels recovered by NIST.	53

LIST OF ACRONYMS AND ABBREVIATIONS

Acronyms

AISC	American Institute of Steel Construction
ASCE	American Society of Civil Engineers
ASTM	ASTM International
BOCA	Building Officials and Code Administrators
BOCA/BBC	BOCA Basic Building Code
DTAP	dissemination and technical assistance program
FEMA	Federal Emergency Management Agency
GMS, LLP	Gilsanz Murray Steficek, LLP
JFK	John F. Kennedy International Airport
LERA	Leslie E. Robertson Associates
LES	Large Eddy Simulation
NIST	National Institute of Standards and Technology
P.L.	Public Law
PANYNJ	Port Authority of New York and New Jersey
PONYA	Port of New York Authority
R&D	research and development
SEAoNY	Structural Engineers Association of New York
USC	United States Code
WF	wide flange (a type of structural steel shape now usually called a W-shape). ASTM A 6 defines them as "doubly-symmetric, wide-flange shapes with inside flange surfaces that are substantially parallel."
WTC	World Trade Center
WTC 1	World Trade Center 1 (North Tower)
WTC 2	World Trade Center 2 (South Tower)
WTC 7	World Trade Center 7

Abbreviations

ft	foot
F_y	yield strength (AISC usage)
in.	inch
kg	kilogram
ksi	1,000 pounds per square inch
m	meter

PREFACE

Genesis of This Investigation

Immediately following the terrorist attack on the World Trade Center (WTC) on September 11, 2001, the Federal Emergency Management Agency (FEMA) and the American Society of Civil Engineers began planning a building performance study of the disaster. The week of October 7, as soon as the rescue and search efforts ceased, the Building Performance Study Team went to the site and began its assessment. This was to be a brief effort, as the study team consisted of experts who largely volunteered their time away from their other professional commitments. The Building Performance Study Team issued its report in May 2002, fulfilling its goal "to determine probable failure mechanisms and to identify areas of future investigation that could lead to practical measures for improving the damage resistance of buildings against such unforeseen events."

On August 21, 2002, with funding from the U.S. Congress through FEMA, the National Institute of Standards and Technology (NIST) announced its building and fire safety investigation of the WTC disaster. On October 1, 2002, the National Construction Safety Team Act (Public Law 107-231), was signed into law. The NIST WTC Investigation was conducted under the authority of the National Construction Safety Team Act.

The goals of the investigation of the WTC disaster were:

- To investigate the building construction, the materials used, and the technical conditions that contributed to the outcome of the WTC disaster.
- To serve as the basis for:
  - Improvements in the way buildings are designed, constructed, maintained, and used;
  - Improved tools and guidance for industry and safety officials;
  - Recommended revisions to current codes, standards, and practices; and
  - Improved public safety.

The specific objectives were:

1. Determine why and how WTC 1 and WTC 2 collapsed following the initial impacts of the aircraft and why and how WTC 7 collapsed;
2. Determine why the injuries and fatalities were so high or low depending on location, including all technical aspects of fire protection, occupant behavior, evacuation, and emergency response;
3. Determine what procedures and practices were used in the design, construction, operation, and maintenance of WTC 1, 2, and 7; and
4. Identify, as specifically as possible, areas in current building and fire codes, standards, and practices that warrant revision.

NIST is a nonregulatory agency of the U.S. Department of Commerce's Technology Administration. The purpose of NIST investigations is to improve the safety and structural integrity of buildings in the United States, and the focus is on fact finding. NIST investigative teams are authorized to assess building performance and emergency response and evacuation procedures in the wake of any building failure that has resulted in substantial loss of life or that posed significant potential of substantial loss of life. NIST does not have the statutory authority to make findings of fault nor negligence by individuals or organizations. Further, no part of any report resulting from a NIST investigation into a building failure or from an investigation under the National Construction Safety Team Act may be used in any suit or action for damages arising out of any matter mentioned in such report (15 USC 281a, as amended by Public Law 107-231).

Organization of the Investigation

The National Construction Safety Team for this Investigation, appointed by the then NIST Director, Dr. Arden L. Bement, Jr., was led by Dr. S. Shyam Sunder. Dr. William L. Grosshandler served as Associate Lead Investigator, Mr. Stephen A. Cauffman served as Program Manager for Administration, and Mr. Harold E. Nelson served on the team as a private sector expert. The Investigation included eight interdependent projects whose leaders comprised the remainder of the team. A detailed description of each of these eight projects is available at http://wtc.nist.gov. The purpose of each project is summarized in Table P–1, and the key interdependencies among the projects are illustrated in Fig. P–1.

Table P–1. Federal building and fire safety investigation of the WTC disaster.

Technical Area and Project Leader	Project Purpose
Analysis of Building and Fire Codes and Practices; Project Leaders: Dr. H. S. Lew and Mr. Richard W. Bukowski	Document and analyze the code provisions, procedures, and practices used in the design, construction, operation, and maintenance of the structural, passive fire protection, and emergency access and evacuation systems of WTC 1, 2, and 7.
Baseline Structural Performance and Aircraft Impact Damage Analysis; Project Leader: Dr. Fahim H. Sadek	Analyze the baseline performance of WTC 1 and WTC 2 under design, service, and abnormal loads, and aircraft impact damage on the structural, fire protection, and egress systems.
Mechanical and Metallurgical Analysis of Structural Steel; Project Leader: Dr. Frank W. Gayle	Determine and analyze the mechanical and metallurgical properties and quality of steel, weldments, and connections from steel recovered from WTC 1, 2, and 7.
Investigation of Active Fire Protection Systems; Project Leader: Dr. David D. Evans; Dr. William Grosshandler	Investigate the performance of the active fire protection systems in WTC 1, 2, and 7 and their role in fire control, emergency response, and fate of occupants and responders.
Reconstruction of Thermal and Tenability Environment; Project Leader: Dr. Richard G. Gann	Reconstruct the time-evolving temperature, thermal environment, and smoke movement in WTC 1, 2, and 7 for use in evaluating the structural performance of the buildings and behavior and fate of occupants and responders.
Structural Fire Response and Collapse Analysis; Project Leaders: Dr. John L. Gross and Dr. Therese P. McAllister	Analyze the response of the WTC towers to fires with and without aircraft damage, the response of WTC 7 in fires, the performance of composite steel-trussed floor systems, and determine the most probable structural collapse sequence for WTC 1, 2, and 7.
Occupant Behavior, Egress, and Emergency Communications; Project Leader: Mr. Jason D. Averill	Analyze the behavior and fate of occupants and responders, both those who survived and those who did not, and the performance of the evacuation system.
Emergency Response Technologies and Guidelines; Project Leader: Mr. J. Randall Lawson	Document the activities of the emergency responders from the time of the terrorist attacks on WTC 1 and WTC 2 until the collapse of WTC 7, including practices followed and technologies used.

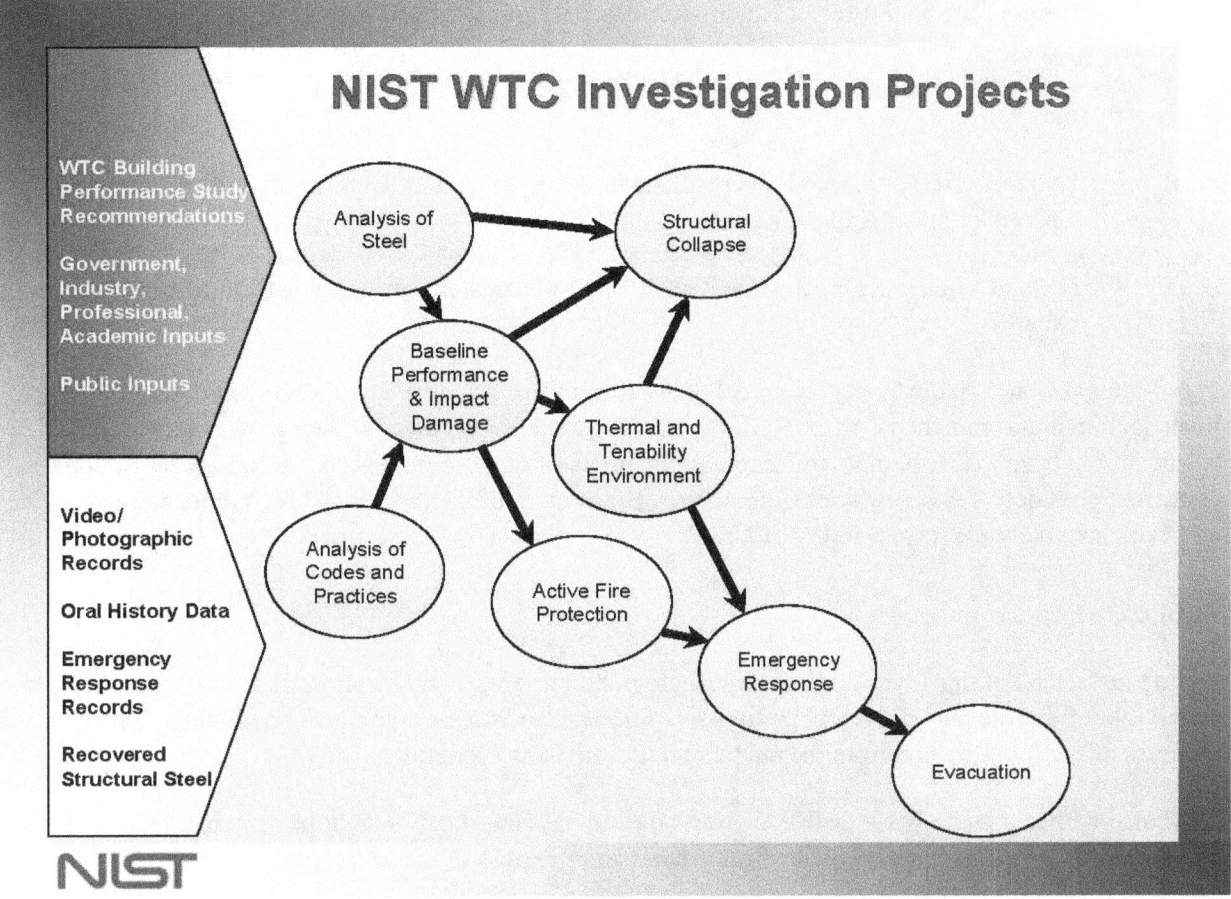

Figure P–1. The eight projects in the federal building and fire safety investigation of the WTC disaster.

National Construction Safety Team Advisory Committee

The NIST Director also established an advisory committee as mandated under the National Construction Safety Team Act. The initial members of the committee were appointed following a public solicitation. These were:

- Paul Fitzgerald, Executive Vice President (retired) FM Global, National Construction Safety Team Advisory Committee Chair

- John Barsom, President, Barsom Consulting, Ltd.

- John Bryan, Professor Emeritus, University of Maryland

- David Collins, President, The Preview Group, Inc.

- Glenn Corbett, Professor, John Jay College of Criminal Justice

- Philip DiNenno, President, Hughes Associates, Inc.

- Robert Hanson, Professor Emeritus, University of Michigan

- Charles Thornton, Co-Chairman and Managing Principal, The Thornton-Tomasetti Group, Inc.

- Kathleen Tierney, Director, Natural Hazards Research and Applications Information Center, University of Colorado at Boulder

- Forman Williams, Director, Center for Energy Research, University of California at San Diego

This National Construction Safety Team Advisory Committee provided technical advice during the Investigation and commentary on drafts of the Investigation reports prior to their public release. NIST has benefited from the work of many people in the preparation of these reports, including the National Construction Safety Team Advisory Committee. The content of the reports and recommendations, however, are solely the responsibility of NIST.

Public Outreach

During the course of this Investigation, NIST held public briefings and meetings (listed in Table P–2) to solicit input from the public, present preliminary findings, and obtain comments on the direction and progress of the Investigation from the public and the Advisory Committee.

NIST maintained a publicly accessible Web site during this Investigation at http://wtc.nist.gov. The site contained extensive information on the background and progress of the Investigation.

NIST's WTC Public-Private Response Plan

The collapse of the WTC buildings has led to broad reexamination of how tall buildings are designed, constructed, maintained, and used, especially with regard to major events such as fires, natural disasters, and terrorist attacks. Reflecting the enhanced interest in effecting necessary change, NIST, with support from Congress and the Administration, has put in place a program, the goal of which is to develop and implement the standards, technology, and practices needed for cost-effective improvements to the safety and security of buildings and building occupants, including evacuation, emergency response procedures, and threat mitigation.

The strategy to meet this goal is a three-part NIST-led public-private response program that includes:

- A federal building and fire safety investigation to study the most probable factors that contributed to post-aircraft impact collapse of the WTC towers and the 47-story WTC 7 building, and the associated evacuation and emergency response experience.

- A research and development (R&D) program to (a) facilitate the implementation of recommendations resulting from the WTC Investigation, and (b) provide the technical basis for cost-effective improvements to national building and fire codes, standards, and practices that enhance the safety of buildings, their occupants, and emergency responders.

Table P–2. Public meetings and briefings of the WTC Investigation.

Date	Location	Principal Agenda
June 24, 2002	New York City, NY	Public meeting: Public comments on the *Draft Plan* for the pending WTC Investigation.
August 21, 2002	Gaithersburg, MD	Media briefing announcing the formal start of the Investigation.
December 9, 2002	Washington, DC	Media briefing on release of the *Public Update* and NIST request for photographs and videos.
April 8, 2003	New York City, NY	Joint public forum with Columbia University on first-person interviews.
April 29–30, 2003	Gaithersburg, MD	NCST Advisory Committee meeting on plan for and progress on WTC Investigation with a public comment session.
May 7, 2003	New York City, NY	Media briefing on release of *May 2003 Progress Report*.
August 26–27, 2003	Gaithersburg, MD	NCST Advisory Committee meeting on status of the WTC investigation with a public comment session.
September 17, 2003	New York City, NY	Media and public briefing on initiation of first-person data collection projects.
December 2–3, 2003	Gaithersburg, MD	NCST Advisory Committee meeting on status and initial results and release of the *Public Update* with a public comment session.
February 12, 2004	New York City, NY	Public meeting on progress and preliminary findings with public comments on issues to be considered in formulating final recommendations.
June 18, 2004	New York City, NY	Media/public briefing on release of *June 2004 Progress Report*.
June 22–23, 2004	Gaithersburg, MD	NCST Advisory Committee meeting on the status of and preliminary findings from the WTC Investigation with a public comment session.
August 24, 2004	Northbrook, IL	Public viewing of standard fire resistance test of WTC floor system at Underwriters Laboratories, Inc.
October 19–20, 2004	Gaithersburg, MD	NCST Advisory Committee meeting on status and near complete set of preliminary findings with a public comment session.
November 22, 2004	Gaithersburg, MD	NCST Advisory Committee discussion on draft annual report to Congress, a public comment session, and a closed session to discuss pre-draft recommendations for WTC Investigation.
April 5, 2005	New York City, NY	Media and public briefing on release of the probable collapse sequence for the WTC towers and draft reports for the projects on codes and practices, evacuation, and emergency response.
June 23, 2005	New York City, NY	Media and public briefing on release of all draft reports for the WTC towers and draft recommendations for public comment.
September 12–13, 2005	Gaithersburg, MD	NCST Advisory Committee meeting on disposition of public comments and update to draft reports for the WTC towers.
September 13–15, 2005	Gaithersburg, MD	WTC Technical Conference for stakeholders and technical community for dissemination of findings and recommendations and opportunity for public to make technical comments.

- A dissemination and technical assistance program (DTAP) to (a) engage leaders of the construction and building community in ensuring timely adoption and widespread use of proposed changes to practices, standards, and codes resulting from the WTC Investigation and the R&D program, and (b) provide practical guidance and tools to better prepare facility owners, contractors, architects, engineers, emergency responders, and regulatory authorities to respond to future disasters.

The desired outcomes are to make buildings, occupants, and first responders safer in future disaster events.

Preface

National Construction Safety Team Reports on the WTC Investigation

A final report on the collapse of the WTC towers is being issued as NIST NCSTAR 1. A companion report on the collapse of WTC 7 is being issued as NIST NCSTAR 1A. The present report is one of a set that provides more detailed documentation of the Investigation findings and the means by which these technical results were achieved. As such, it is part of the archival record of this Investigation. The titles of the full set of Investigation publications are:

NIST (National Institute of Standards and Technology). 2005. *Federal Building and Fire Safety Investigation of the World Trade Center Disaster: Final Report on the Collapse of the World Trade Center Towers.* NIST NCSTAR 1. Gaithersburg, MD, September.

NIST (National Institute of Standards and Technology). 2008. *Federal Building and Fire Safety Investigation of the World Trade Center Disaster: Final Report on the Collapse of World Trade Center 7.* NIST NCSTAR 1A. Gaithersburg, MD, November.

Lew, H. S., R. W. Bukowski, and N. J. Carino. 2005. *Federal Building and Fire Safety Investigation of the World Trade Center Disaster: Design, Construction, and Maintenance of Structural and Life Safety Systems.* NIST NCSTAR 1-1. National Institute of Standards and Technology. Gaithersburg, MD, September.

> Fanella, D. A., A. T. Derecho, and S. K. Ghosh. 2005. *Federal Building and Fire Safety Investigation of the World Trade Center Disaster: Design and Construction of Structural Systems.* NIST NCSTAR 1-1A. National Institute of Standards and Technology. Gaithersburg, MD, September.
>
> Ghosh, S. K., and X. Liang. 2005. *Federal Building and Fire Safety Investigation of the World Trade Center Disaster: Comparison of Building Code Structural Requirements.* NIST NCSTAR 1-1B. National Institute of Standards and Technology. Gaithersburg, MD, September.
>
> Fanella, D. A., A. T. Derecho, and S. K. Ghosh. 2005. *Federal Building and Fire Safety Investigation of the World Trade Center Disaster: Maintenance and Modifications to Structural Systems.* NIST NCSTAR 1-1C. National Institute of Standards and Technology. Gaithersburg, MD, September.
>
> Grill, R. A., and D. A. Johnson. 2005. *Federal Building and Fire Safety Investigation of the World Trade Center Disaster: Fire Protection and Life Safety Provisions Applied to the Design and Construction of World Trade Center 1, 2, and 7 and Post-Construction Provisions Applied after Occupancy.* NIST NCSTAR 1-1D. National Institute of Standards and Technology. Gaithersburg, MD, September.
>
> Razza, J. C., and R. A. Grill. 2005. *Federal Building and Fire Safety Investigation of the World Trade Center Disaster: Comparison of Codes, Standards, and Practices in Use at the Time of the Design and Construction of World Trade Center 1, 2, and 7.* NIST NCSTAR 1-1E. National Institute of Standards and Technology. Gaithersburg, MD, September.
>
> Grill, R. A., D. A. Johnson, and D. A. Fanella. 2005. *Federal Building and Fire Safety Investigation of the World Trade Center Disaster: Comparison of the 1968 and Current (2003) New*

York City Building Code Provisions. NIST NCSTAR 1-1F. National Institute of Standards and Technology. Gaithersburg, MD, September.

Grill, R. A., and D. A. Johnson. 2005. *Federal Building and Fire Safety Investigation of the World Trade Center Disaster: Amendments to the Fire Protection and Life Safety Provisions of the New York City Building Code by Local Laws Adopted While World Trade Center 1, 2, and 7 Were in Use*. NIST NCSTAR 1-1G. National Institute of Standards and Technology. Gaithersburg, MD, September.

Grill, R. A., and D. A. Johnson. 2005. *Federal Building and Fire Safety Investigation of the World Trade Center Disaster: Post-Construction Modifications to Fire Protection and Life Safety Systems of World Trade Center 1 and 2*. NIST NCSTAR 1-1H. National Institute of Standards and Technology. Gaithersburg, MD, September.

Grill, R. A., D. A. Johnson, and D. A. Fanella. 2005. *Federal Building and Fire Safety Investigation of the World Trade Center Disaster: Post-Construction Modifications to Fire Protection, Life Safety, and Structural Systems of World Trade Center 7*. NIST NCSTAR 1-1I. National Institute of Standards and Technology. Gaithersburg, MD, September.

Grill, R. A., and D. A. Johnson. 2005. *Federal Building and Fire Safety Investigation of the World Trade Center Disaster: Design, Installation, and Operation of Fuel System for Emergency Power in World Trade Center 7*. NIST NCSTAR 1-1J. National Institute of Standards and Technology. Gaithersburg, MD, September.

Sadek, F. 2005. *Federal Building and Fire Safety Investigation of the World Trade Center Disaster: Baseline Structural Performance and Aircraft Impact Damage Analysis of the World Trade Center Towers*. NIST NCSTAR 1-2. National Institute of Standards and Technology. Gaithersburg, MD, September.

Faschan, W. J., and R. B. Garlock. 2005. *Federal Building and Fire Safety Investigation of the World Trade Center Disaster: Reference Structural Models and Baseline Performance Analysis of the World Trade Center Towers*. NIST NCSTAR 1-2A. National Institute of Standards and Technology. Gaithersburg, MD, September.

Kirkpatrick, S. W., R. T. Bocchieri, F. Sadek, R. A. MacNeill, S. Holmes, B. D. Peterson, R. W. Cilke, C. Navarro. 2005. *Federal Building and Fire Safety Investigation of the World Trade Center Disaster: Analysis of Aircraft Impacts into the World Trade Center Towers*, NIST NCSTAR 1-2B. National Institute of Standards and Technology. Gaithersburg, MD, September.

Gayle, F. W., R. J. Fields, W. E. Luecke, S. W. Banovic, T. Foecke, C. N. McCowan, T. A. Siewert, and J. D. McColskey. 2005. *Federal Building and Fire Safety Investigation of the World Trade Center Disaster: Mechanical and Metallurgical Analysis of Structural Steel*. NIST NCSTAR 1-3. National Institute of Standards and Technology. Gaithersburg, MD, September.

Luecke, W. E., T. A. Siewert, and F. W. Gayle. 2005. *Federal Building and Fire Safety Investigation of the World Trade Center Disaster: Contemporaneous Structural Steel Specifications*. NIST Special Publication 1-3A. National Institute of Standards and Technology. Gaithersburg, MD, September.

Banovic, S. W. 2005. *Federal Building and Fire Safety Investigation of the World Trade Center Disaster: Steel Inventory and Identification.* NIST NCSTAR 1-3B. National Institute of Standards and Technology. Gaithersburg, MD, September.

Banovic, S. W., and T. Foecke. 2005. *Federal Building and Fire Safety Investigation of the World Trade Center Disaster: Damage and Failure Modes of Structural Steel Components.* NIST NCSTAR 1-3C. National Institute of Standards and Technology. Gaithersburg, MD, September.

Luecke, W. E., J. D. McColskey, C. N. McCowan, S. W. Banovic, R. J. Fields, T. Foecke, T. A. Siewert, and F. W. Gayle. 2005. *Federal Building and Fire Safety Investigation of the World Trade Center Disaster: Mechanical Properties of Structural Steels.* NIST NCSTAR 1-3D. National Institute of Standards and Technology. Gaithersburg, MD, September.

Banovic, S. W., C. N. McCowan, and W. E. Luecke. 2005. *Federal Building and Fire Safety Investigation of the World Trade Center Disaster: Physical Properties of Structural Steels.* NIST NCSTAR 1-3E. National Institute of Standards and Technology. Gaithersburg, MD, September.

Evans, D. D., R. D. Peacock, E. D. Kuligowski, W. S. Dols, and W. L. Grosshandler. 2005. *Federal Building and Fire Safety Investigation of the World Trade Center Disaster: Active Fire Protection Systems.* NIST NCSTAR 1-4. National Institute of Standards and Technology. Gaithersburg, MD, September.

Kuligowski, E. D., D. D. Evans, and R. D. Peacock. 2005. *Federal Building and Fire Safety Investigation of the World Trade Center Disaster: Post-Construction Fires Prior to September 11, 2001.* NIST NCSTAR 1-4A. National Institute of Standards and Technology. Gaithersburg, MD, September.

Hopkins, M., J. Schoenrock, and E. Budnick. 2005. *Federal Building and Fire Safety Investigation of the World Trade Center Disaster: Fire Suppression Systems.* NIST NCSTAR 1-4B. National Institute of Standards and Technology. Gaithersburg, MD, September.

Keough, R. J., and R. A. Grill. 2005. *Federal Building and Fire Safety Investigation of the World Trade Center Disaster: Fire Alarm Systems.* NIST NCSTAR 1-4C. National Institute of Standards and Technology. Gaithersburg, MD, September.

Ferreira, M. J., and S. M. Strege. 2005. *Federal Building and Fire Safety Investigation of the World Trade Center Disaster: Smoke Management Systems.* NIST NCSTAR 1-4D. National Institute of Standards and Technology. Gaithersburg, MD, September.

Gann, R. G., A. Hamins, K. B. McGrattan, G. W. Mulholland, H. E. Nelson, T. J. Ohlemiller, W. M. Pitts, and K. R. Prasad. 2005. *Federal Building and Fire Safety Investigation of the World Trade Center Disaster: Reconstruction of the Fires in the World Trade Center Towers.* NIST NCSTAR 1-5. National Institute of Standards and Technology. Gaithersburg, MD, September.

Pitts, W. M., K. M. Butler, and V. Junker. 2005. *Federal Building and Fire Safety Investigation of the World Trade Center Disaster: Visual Evidence, Damage Estimates, and Timeline Analysis.* NIST NCSTAR 1-5A. National Institute of Standards and Technology. Gaithersburg, MD, September.

Hamins, A., A. Maranghides, K. B. McGrattan, E. Johnsson, T. J. Ohlemiller, M. Donnelly, J. Yang, G. Mulholland, K. R. Prasad, S. Kukuck, R. Anleitner and T. McAllister. 2005. *Federal Building and Fire Safety Investigation of the World Trade Center Disaster: Experiments and Modeling of Structural Steel Elements Exposed to Fire.* NIST NCSTAR 1-5B. National Institute of Standards and Technology. Gaithersburg, MD, September.

Ohlemiller, T. J., G. W. Mulholland, A. Maranghides, J. J. Filliben, and R. G. Gann. 2005. *Federal Building and Fire Safety Investigation of the World Trade Center Disaster: Fire Tests of Single Office Workstations.* NIST NCSTAR 1-5C. National Institute of Standards and Technology. Gaithersburg, MD, September.

Gann, R. G., M. A. Riley, J. M. Repp, A. S. Whittaker, A. M. Reinhorn, and P. A. Hough. 2005. *Federal Building and Fire Safety Investigation of the World Trade Center Disaster: Reaction of Ceiling Tile Systems to Shocks.* NIST NCSTAR 1-5D. National Institute of Standards and Technology. Gaithersburg, MD, September.

Hamins, A., A. Maranghides, K. B. McGrattan, T. J. Ohlemiller, and R. Anleitner. 2005. *Federal Building and Fire Safety Investigation of the World Trade Center Disaster: Experiments and Modeling of Multiple Workstations Burning in a Compartment.* NIST NCSTAR 1-5E. National Institute of Standards and Technology. Gaithersburg, MD, September.

McGrattan, K. B., C. Bouldin, and G. Forney. 2005. *Federal Building and Fire Safety Investigation of the World Trade Center Disaster: Computer Simulation of the Fires in the World Trade Center Towers.* NIST NCSTAR 1-5F. National Institute of Standards and Technology. Gaithersburg, MD, September.

Prasad, K. R., and H. R. Baum. 2005. *Federal Building and Fire Safety Investigation of the World Trade Center Disaster: Fire Structure Interface and Thermal Response of the World Trade Center Towers.* NIST NCSTAR 1-5G. National Institute of Standards and Technology. Gaithersburg, MD, September.

Gross, J. L., and T. McAllister. 2005. *Federal Building and Fire Safety Investigation of the World Trade Center Disaster: Structural Fire Response and Probable Collapse Sequence of the World Trade Center Towers.* NIST NCSTAR 1-6. National Institute of Standards and Technology. Gaithersburg, MD, September.

Carino, N. J., M. A. Starnes, J. L. Gross, J. C. Yang, S. Kukuck, K. R. Prasad, and R. W. Bukowski. 2005. *Federal Building and Fire Safety Investigation of the World Trade Center Disaster: Passive Fire Protection.* NIST NCSTAR 1-6A. National Institute of Standards and Technology. Gaithersburg, MD, September.

Gross, J., F. Hervey, M. Izydorek, J. Mammoser, and J. Treadway. 2005. *Federal Building and Fire Safety Investigation of the World Trade Center Disaster: Fire Resistance Tests of Floor Truss Systems.* NIST NCSTAR 1-6B. National Institute of Standards and Technology. Gaithersburg, MD, September.

Zarghamee, M. S., S. Bolourchi, D. W. Eggers, Ö. O. Erbay, F. W. Kan, Y. Kitane, A. A. Liepins, M. Mudlock, W. I. Naguib, R. P. Ojdrovic, A. T. Sarawit, P. R Barrett, J. L. Gross, and

T. P. McAllister. 2005. *Federal Building and Fire Safety Investigation of the World Trade Center Disaster: Component, Connection, and Subsystem Structural Analysis.* NIST NCSTAR 1-6C. National Institute of Standards and Technology. Gaithersburg, MD, September.

Zarghamee, M. S., Y. Kitane, Ö. O. Erbay, T. P. McAllister, and J. L. Gross. 2005. *Federal Building and Fire Safety Investigation of the World Trade Center Disaster: Global Structural Analysis of the Response of the World Trade Center Towers to Impact Damage and Fire.* NIST NCSTAR 1-6D. National Institute of Standards and Technology. Gaithersburg, MD, September.

Averill, J. D., D. S. Mileti, R. D. Peacock, E. D. Kuligowski, N. Groner, G. Proulx, P. A. Reneke, and H. E. Nelson. 2005. *Federal Building and Fire Safety Investigation of the World Trade Center Disaster: Occupant Behavior, Egress, and Emergency Communication.* NIST NCSTAR 1-7. National Institute of Standards and Technology. Gaithersburg, MD, September.

Fahy, R., and G. Proulx. 2005. *Federal Building and Fire Safety Investigation of the World Trade Center Disaster: Analysis of Published Accounts of the World Trade Center Evacuation.* NIST NCSTAR 1-7A. National Institute of Standards and Technology. Gaithersburg, MD, September.

Zmud, J. 2005. *Federal Building and Fire Safety Investigation of the World Trade Center Disaster: Technical Documentation for Survey Administration.* NIST NCSTAR 1-7B. National Institute of Standards and Technology. Gaithersburg, MD, September.

Lawson, J. R., and R. L. Vettori. 2005. *Federal Building and Fire Safety Investigation of the World Trade Center Disaster: The Emergency Response Operations.* NIST NCSTAR 1-8. National Institute of Standards and Technology. Gaithersburg, MD, September.

McAllister, T., R. G. Gann, J. D. Averill, J. L. Gross, W. L. Grosshandler, J. R. Lawson, K. B. McGrattan, H. E. Nelson, W. M. Pitts, K. R. Prasad, F. H. Sadek. 2008. *Federal Building and Fire Safety Investigation of the World Trade Center Disaster: Structural Fire Response and Probable Collapse Sequence of World Trade Center Building 7.* NIST NCSTAR 1-9. National Institute of Standards and Technology. Gaithersburg, MD, November.

MacNeill, R., S. Kirkpatrick, B. Peterson, and R. Bocchieri. 2008. *Federal Building and Fire Safety Investigation of the World Trade Center Disaster: Global Structural Analysis of the Response of World Trade Center Building 7 to Fires and Debris Impact Damage.* NIST NCSTAR 1-9A. National Institute of Standards and Technology. Gaithersburg, MD, November.

ACKNOWLEDGMENTS

National Institute of Standards and Technology (NIST) thanks these volunteers of the Structural Engineers Association of New York for their efforts in the recovery of the steel components: Amit Bandyopadhyay, Anamaria Bonilla, Peter Chipchase, Anthony Chuliver, Edward DePaola, Louis Errichiello, James Fahey, Ramon Gilsanz, Jeffrey Hartman, David Hoy, Dean Koutsoubis, Andrew McConnell, Rajani Nair, Alan Rosa, David Sharp, Gary Steficek, and Kevin Terry. Countless hours were unselfishly spent in the recovery yards searching for these invaluable pieces that are an integral component of this investigation. Further, the following people are recognized for their outstanding leadership roles throughout the recovery effort: Ramon Gilsanz, primary leader of the recovery effort; David Sharp, coordinator of volunteer activities and author of the guide used for the selection of suitable pieces; and Audrey Massa of Federal Emergency Management Agency (FEMA), leader for documentation and cataloguing of efforts and pieces to be saved.

The FEMA/American Society of Civil Engineers Building Performance Assessment Team; Professor A. Astaneh-Asl of the University of California, Berkeley, California; and the National Science Foundation are also acknowledged for their help in the recovery effort.

Blanford Land Development Corporation; Hugo Neu Schnitzer, Inc.; and Metal Management, Inc. are thanked for the initial storing of the structural steel during the clean-up effort of the World Trade Center site and for their assistance and patience during the review, abatement, and final removal of pieces of interest to the investigation.

NIST also thanks the Port Authority of New York and New Jersey for its assistance in surveying, sectioning, and transporting the technically significant structural steel collected by the Port Authority and stored at John F. Kennedy International Airport.

EXECUTIVE SUMMARY

As a result of the recovery efforts of the Structural Engineers Association of New York, Federal Emergency Management Agency/American Society of Civil Engineers, and the National Institute of Standards and Technology (NIST), NIST possesses 236 structural steel elements from the World Trade Center (WTC) buildings. These pieces represent a small fraction of the enormous amount of steel examined at the various recovery yards where the debris was sent as the WTC site was cleared. Elements located in or near the impact zone and fire damaged regions were emphasized in the selection process. These samples include full exterior column panels, core columns, portions of the floor truss members, channels used to attach the floor trusses to the interior columns, and other smaller structural components (e.g., bolts, diagonal bracing straps, aluminum façade, etc.). These structural components were used for evaluation and/or testing relative to the fire and structural response of the WTC buildings.

Upon arrival at NIST, the samples were catalogued, documented, and when possible, identified as to their precise, as-built location within the buildings. The vast majority of the structural components are from WTC 1 and WTC 2. It is estimated that roughly 0.25 percent to 0.5 percent of the 200,000 tons of steel used in the construction of the two towers was recovered. The following lists the recovered structural steel elements:

- Out of the 90 exterior panels recovered, the as-built location of 42 distinct sections was unambiguously identified within WTC 1 and WTC 2:

  - 26 panels from WTC 1: 22 from or near the impact floors, 4 hit directly by the airplane,

  - 16 panels from WTC 2: 4 near the impact floors.

- Out of the 55 wide flange sections and built-up box sections recovered, 12 core columns were positively identified from WTC 1 and WTC 2, including:

  - Two columns from the fire floors of WTC 1,

  - Two columns from the impact zone of WTC 2.

- 23 pieces of floor truss material from WTC 1 and WTC 2 were recovered; however, the as-built location of the trusses within the buildings could not be identified.

- 25 pieces of channel material that connected the floor trusses to the core columns in WTC 1 and WTC 2 were recovered; however, the as-built location of the channels within the buildings could not be identified.

- One piece of floor framing from outside the core of the 107th floor of WTC 1.

- Seven coupons from WTC 5 were removed in the field and sent to NIST.

- No pieces could be unambiguously identified as being from WTC 7.

Executive Summary

The design drawings for WTC 1 and WTC 2 designate 14 different grades (or strengths) of steel for the exterior panels, four different grades for the core columns, and two grades for the floor trusses. From the recovered and identified columns, whether perimeter or core, a one to one correlation was observed between the minimum yield strength specified by the design drawings and the observed stampings and/or stencilings on the samples, with the exception of the 85 ksi and 90 ksi material that was substituted with 100 ksi plate. The recovered structural elements yielded sufficient representative samples for the following:

- All 12 grades of exterior panel material,

- Two grades of the core column material (representing 99 percent, by total number, of the columns),

- Both grades for the floor truss material.

This collection of steel from the WTC towers is sufficient for determining the quality of the steel and for determining mechanical properties as input to models of building performance. The lack of WTC 7 steel precludes tests on actual material from the structure; however, WTC 7 was constructed of three grades of conventional steel (36 ksi, 42 ksi, and 50 ksi), and literature values may be used to estimate properties.

Chapter 1
INTRODUCTION

1.1 PURPOSE OF REPORT

The purpose of analyzing structural steel available from World Trade Center (WTC) 1, 2, and 7 is to determine the metallurgical and mechanical properties and quality of the metal, weldments, and connections and to provide these data for other analyses in the National Institute of Standards and Technology (NIST) Investigation. The properties determined were used in two ways:

- Properties were correlated with the design requirements of the buildings to determine if the specified steel was in place in the towers.

- Properties were supplied as input for models of building performance.

1.2 SCOPE OF REPORT

The scope of this report covers the inventory and identification of steels recovered from the WTC buildings. Approximately 236 pieces of WTC steel were available for study at NIST. These pieces represent a small fraction of the steel examined at the various recovery yards where the steel was sent as the WTC site was cleared.

For reference throughout this report, NIST NCSTAR 1-3A[1] describes the tower structure and critical structural elements discussed below. This includes the structural design and properties specified by the structural engineers for columns, floor systems, and connections. Further, NIST NCSTAR 1-3A also discusses the contemporaneous (late 1960s era) specifications for various types and grades of steel designated by ASTM International, the American Institute of Steel Construction (AISC), and other national and international organizations. It also includes information from numerous suppliers of the steel for the structure.

---

[1] This reference is to one of the companion documents from this Investigation. A list of these documents appears in the Preface to this report.

Chapter 2
BACKGROUND INFORMATION RELATED TO RECOVERY OF WTC STRUCTURAL STEEL

Beginning in October 2001, members of the Federal Emergency Management Agency (FEMA), American Society of Civil Engineers (ASCE), Building Performance Study (BPS) Team, members of the Structural Engineers Association of New York (SEAoNY), and Professor A. Astaneh-Asl of the University of California, Berkeley, California (Astaneh-Asl 2002), with support from the National Science Foundation, began work to identify and collect World Trade Center (WTC) structural steel from the various recovery yards where debris, including the steel, was taken during the cleanup effort. Dr. J. Gross, a structural engineer at the National Institute of Standards and Technology (NIST) and a member of the FEMA/ASCE BPS Team, was involved in these early efforts.

There were four major sites where debris from the WTC buildings was shipped during the clean-up effort in which the volunteers worked. These were:

- Hugo Neu Schnitzer, Inc., Fresh Kills Landfill in Staten Island, New Jersey;
- Hugo Neu Schnitzer East, Inc., Claremont Terminal in Jersey City, New Jersey;
- Metal Management, Inc., in Newark, New Jersey; and
- Blanford and Co. in Keasbey, New Jersey.

The volunteers searched through unsorted piles of steel and other debris for pieces from the WTC buildings, specifically searching for (McAllister 2002):

- Exterior column panels and interior core columns from WTC 1 and WTC 2 that were exposed to fire and/or impacted by the aircraft;
- Exterior column panels and interior core columns from WTC 1 and WTC 2 directly above and below the impact zones;
- Badly burned pieces from WTC 7;
- Connections from WTC 1, 2, and 7 (e.g., seat connections, single-shear plates, and column splices);
- Bolts in all conditions;
- Floor trusses, including stiffeners, seats, and other components; and
- Any pieces that in the engineers' professional opinion might be useful.

Chapter 2

Once identified for recovery, the samples were marked as "SAVE" and given an alphanumeric code relative to the recovery yard from which they came and an accession number. Some pieces were not saved in their entirety, but instead, small portions were removed, hereafter called coupons. (Coupons were also removed in the field for WTC 5, held at Gilsanz Murray Steficek, LLP [GMS, LLP], and later brought to NIST.)

Facing concern that the identified steel may not be properly preserved in the recovery yards, NIST arranged for the steel to be shipped to its campus in Gaithersburg, Maryland, starting in March 2002. Professor Astaneh-Asl also granted NIST permission to take custody of the steel that he had personally marked. Before the samples were shipped to the NIST campus, environmental testing for asbestos and analysis of the paint for lead was conducted. Volunteers from SEAoNY, with assistance from additional NIST personnel, continued their presence at the recovery yards and identified, catalogued, and shipped steel specimens to NIST through October 2002. The structural components recovered now constitute the material base from which samples are being removed for further evaluation and or testing relative to the fire and structural response of the WTC buildings as part of the WTC Investigation.

Structural steel elements were also collected and held by the Port Authority of New York and New Jersey (PANYNJ) in Hanger 17 located at John F. Kennedy (JFK) International Airport. The main goal of the PANYNJ project was to decontaminate and preserve the steel, as well as other WTC artifacts, for future exhibits and memorials. A complete listing of the pieces held by PANYNJ can be found in the *Preservation and Inventory Report* prepared by Voorsanger and Associates Architects, PC.[2] NIST personnel visited the hanger and identified 12 additional pieces that were considered important to its Investigation. Six of these samples were moved whole to the Gaithersburg campus. The remaining pieces had portions removed and sent to NIST, with the bulk of the structural element remaining at JFK International Airport.

---

[2] Voorsanger and Associates Architects, PC. 2002. *WTC Archives Interim Storage Facility, JFK International Airport: Preservation and Inventory Report, Draft 2.* New York, NY, November.

Chapter 3
STRUCTURAL ELEMENTS RECOVERED FROM THE WTC BUILDINGS

3.1 PRESENT LOCATION AND LABELING OF STRUCTURAL STEEL ELEMENTS

At present, the National Institute of Standards and Technology (NIST) possesses 236 labeled samples from the World Trade Center (WTC) buildings. While the majority of the NIST-held samples reside on the Gaithersburg campus, some samples were shipped to the Boulder, Colorado, campus for mechanical property testing following initial documentation.

As samples were delivered, overall images of the pieces were taken for record-keeping purposes. An example is shown in Fig. 3–1. Samples are identified by their original alphanumeric identification codes assigned by Structural Engineers Association of New York to be consistent with the Federal Emergency Management Agency report. However, there were cases in which two different codes were found on one piece. In these instances, if the pieces were already undergoing documentation procedures, the first code noted was used. Samples that arrived lacking a code were labeled as part of the U series. Additionally, samples brought from Hanger 17 at John F. Kennedy International Airport maintained their "B"-series labels provided in the Voorsanger report.[2]

Appendix A, Table A–1, is a complete list of each sample received, in alphanumeric order, with its classification, a brief description of the component, and the location of the piece on the NIST campus. These samples range from full exterior column panels to pieces of bolts and bags of glass and other debris fragments. The pieces were classified into one of eight categories:

Classification	No. of Pieces	Symbol
Exterior column panel sections (flat wall or corner)	94	C, CC, or Cn
Bowtie pieces	2	BT
Rectangular built-up box column (not perimeter column)	11	RB
Wide flange sections	44	W
Floor trusses	23	J
Channels	25	Ch
Coupons from WTC 5	7	Cn5
Miscellaneous (isolated bolts, floor hanger components, or other)	30	B,H,O

Tables A–2 through A–11 list the pieces separated by type, and Figs. A–1 through A–14 displays characteristic photographs of the various pieces.

Chapter 3

Source: NIST.

Figure 3–1. Characteristic "overall" view of the samples taken for each piece received. Sample shown here is C-14.

3.2 IDENTIFICATION OF WTC STRUCTURAL STEEL ELEMENTS

Information from Leslie E. Robertson Associates (LERA) indicates that all structural steel pieces in WTC 1 and WTC 2 were uniquely identified by stampings (recessed letters and numbers) and/or painted stencils.[3] NIST has been successful in finding these identification markings on many of the perimeter panel sections, core columns, and other wide flange members. Of the 94 pieces of perimeter panel labeled in Table A–1, 90 distinct panels were observed. (The other four pieces of perimeter column had become separated from the main panel during salvage and were subsequently labeled C-13a, C-16a, C-28b, and K-16a.) At this time, of the 90 panels, 42 distinct exterior column panels have been identified and 1 partially identified. Tables 3–1 and 3–2 list these samples, respectively, with Fig. 3–2 showing the relative locations of the identified exterior panels within the top third of the buildings. Significantly more pieces were recovered from WTC 1 than WTC 2. Table 3–3 lists the 12 core columns in NIST's possession that have been positively identified through their stampings. An additional sample, C-83, is also listed in this group. Though no markings were found on the piece, the shape and dimension of this sample are in conformance with the design drawings for core columns and it has a similar appearance to core column C-90. Additionally, there are 13 pieces of wide flange sections that have stampings and/or markings with a different code, Table 3–4. After review of these stampings by staff members from both the Port Authority of New York and New Jersey (PANYNJ) and LERA, no definite correlation between the markings and the contract document member labels was found. It was the opinion of LERA that the stampings/marks are shop marks that may or may not have been indicated in the original shop drawings. LERA believes that it does not have within its possession any shop drawings that are consistent with the members shown. Further, LERA studied the images to determine if the box members could be grouped into a member type typically found in the WTC buildings and was unsuccessful. Therefore, the as-built location of these pieces could not be determined nor could it be confirmed that they were part of the structural steel from the WTC towers.

The positive identification of the structural elements was made possible by deciphering the stampings and/or stencils found on them. During the fabrication process, the exterior panel sections were stamped at the bottom of the center column on the inside face. These stampings indicated the building, center column line number, and floors spanned by the columns. The core columns had stampings placed at the lower end of the component near the connector. The building was typically represented as "A" for WTC 1 and "B" for WTC 2. An example of a stamping found on an exterior column is shown in Fig. 3–3, where the stamping indicates that the piece was from WTC 2, with center column line number 206, spanning floors 83 through 86. Core column material was found to have similar markings (Fig. 3–4). Other stampings have also been found on the flanges of the perimeter columns that indicated the column type (Fig. 3–5 and Table 3–5) as well as the specified minimum yield strength of the column. All of these stampings typically reside within 1 m from the bottom of the column.

---

[3] Faschan, W. 2002. Leslie E. Robertson Associates, New York, NY, personal communication to F. Gayle, Project Leader, World Trade Center Investigation, National Institute of Standards and Technology, Gaithersburg, MD, May 21.

Table 3–1. Identified exterior column panel pieces from WTC 1 and WTC 2.

NIST Name	Type	Description	Bldg.	Column	Floors	Derrick Division
ASCE 2	C	1 full column	WTC 2	330	40–43	NA
B-1024	C	Full panel	WTC 2	154	21–24	NA
B-1043	C	Full panel	WTC 2	406	40–43	NA
B-1044	C	Full panel	WTC 2	409	40–43	NA
C-10	C	Full panel	WTC 1	451	85–88	5x
C-13	CC	Rectangular column with spandrel	WTC 2	200	90–92	569
C-13a	C	Partial of single column	WTC 2	159	90–92	569
C-14	C	1 column, lower 1/3	WTC 2	300	85–87	570
C-18	C	3 columns, bottom 2/3	WTC 2	230	93–96	NA
C-22	C	3 columns, lower 1/2	WTC 1	157	93–96	69
C-24	C	3 columns, upper 1/3	WTC 2	203	74–77	NA
C-25	C	1 column, lower 1/2	WTC 1	206	89–92	69
C-40	C	2 columns, lower 2/3	WTC 1	136	98–101	6x
C-46	C	Nearly full panel	WTC 2	157	68–71	569
C-48	C	Nearly 2 full columns	WTC 2	442	91–94	NA
C-55	C	1 column, lower 1/3	WTC 1	209	94–97	NA
C-89	C	2 full columns	WTC 2	215	12–15	NA
C-92	C	1 column, lower 1/3	WTC 2	130	93–96	NA
C-93	C	1 column, lower 1/3	WTC 1	339	99–102	NA
CC	C	2 full columns	WTC 1	124	70–73	NA
K-1	C	3 columns, lower 1/3	WTC 1	209	97–100	NA
K-2	C	1 column, lower 2/3	WTC 1	236	92–95	NA
M-2	C	Full panel	WTC 1	130	96–99	63
M-10a	C	3 columns, middle section 1/3	WTC 2	209	82–85	NA
M-10b	C	3 columns, lower 1/2	WTC 2	206	83–86	569
M-20	C	2 columns, lower 1/3	WTC 1	121	99–102	63
M-26	C	Full panel	WTC 1	130	90–93	6x
M-27	C	2 columns, lower ¾	WTC 1	130	93–96	63
M-28	C	3 columns, lower ¼	WTC 2	345	98–101	NA
M-30	C	2 columns, lower 1/3	WTC 1	133	94–97	65
N-1	C	2 full columns	WTC 1	218	82–85	NA
N-7	C	Full panel	WTC 1	127	97–100	NA
N-8	C	Full panel	WTC 1	142	97–100	67
N-9	C	Nearly full panel	WTC 1	154	101–104	69
N-10	C	2 columns, lower 2/3	WTC 1	115	89–92	6x
N-12	C	2 full columns	WTC 1	206	92–95	69
N-13	C	3 columns, lower 1/3	WTC 1	130	99–102	63

Table 3–1. Identified exterior column panel pieces from WTC 1 and WTC 2 (continued).

NIST Name	Type	Description	Bldg.	Column	Floors	Derrick Division
N-99	C	Nearly full panel	WTC 1	148	99–102	67
N-101	C	Full panel	WTC 1	133	100–103	65
S-1	C	2 columns, lower 1/3	WTC 1	433	79–82	47
S-9	C	Full panel	WTC 1	133	97–100	NA
S-10	C	2 columns, lower 1/2	WTC 1	224	92–95	NA
S-14	C	Full panel	WTC 2	218	91–94	557

Key: NA, information not available.
Note: "x" in the derrick division column indicates an unreadable number.

Table 3–2. Partially identified exterior column panel from WTC 1 or WTC 2.

NIST Name	Type	Description	Bldg.	Column	Floors
C-117	C	3 columns, lower 1/3	NA	NA	100–104

Key: NA, information not available.

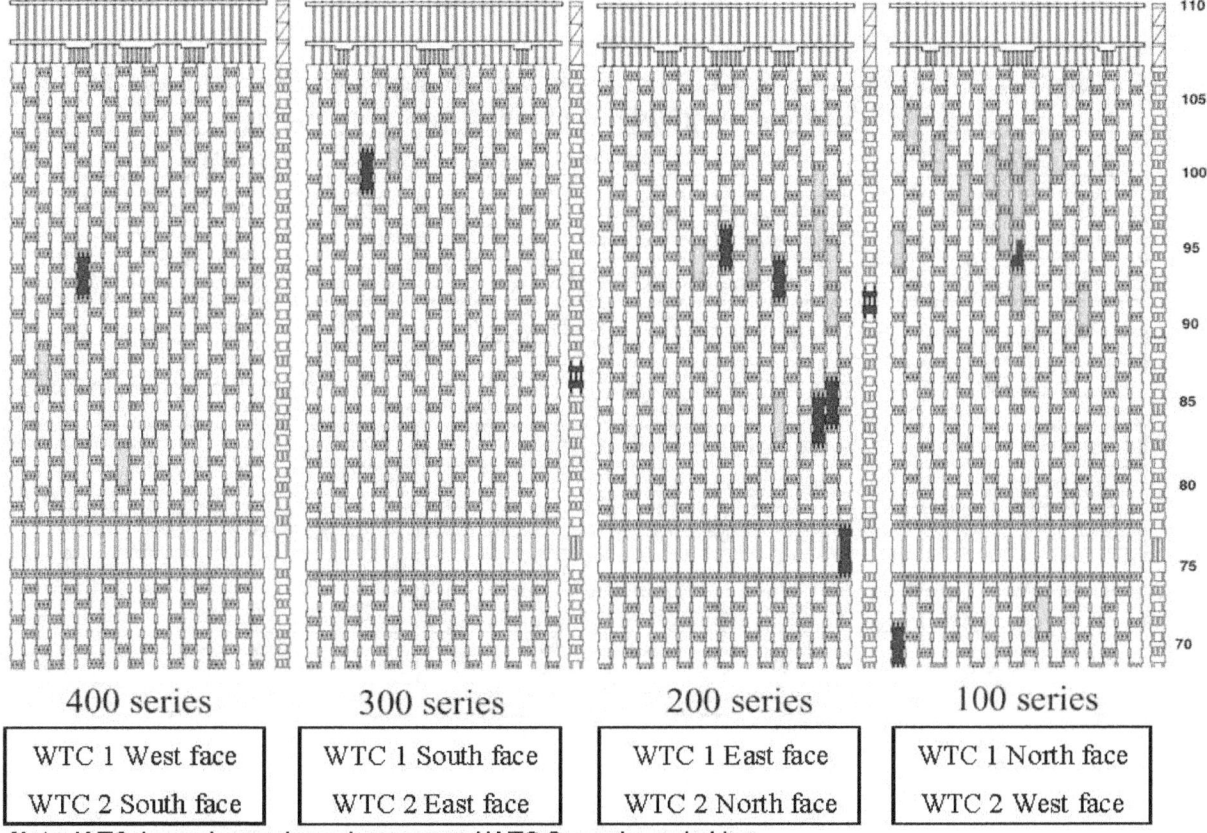

400 series	300 series	200 series	100 series
WTC 1 West face	WTC 1 South face	WTC 1 East face	WTC 1 North face
WTC 2 South face	WTC 2 East face	WTC 2 North face	WTC 2 West face

Note: WTC 1 panels are shown in green and WTC 2 panels are in blue.

Figure 3–2. Location of the exterior panels recovered from the top third of WTC 1 and WTC 2.

Table 3–3. Identified pieces of core column material from WTC 1 and WTC 2.

NIST Name	Type	Description	Bldg.	Column	Floors	Derrick Division	F_y (ksi)
B-1011	RB	Heavy rectangular column	WTC 1	508	51–54	55	36
B-6152-1	RB	Heavy rectangular column	WTC 1	803	15–18	52	36
B-6152-2	RB	Heavy rectangular column	WTC 1	504	33–36	51	36
C-83[a]	RB	Heavy rectangular column	NA	NA	NA	NA	NA
C-88a	RB	Heavy rectangular column	WTC 2	801	80–83	550	42
C-88b	RB	Heavy rectangular column	WTC 2	801	77–80	550	42
C-90	RB	Heavy rectangular column	WTC 2	701	12–15	549	36
C-30 or S-12	W	Wide flange section	WTC 2	1008	104–106	NA	36
C-65 or S-8	W	Wide flange section	WTC 1	904	86–89	52	36
C-71	W	Wide flange section	WTC 1	904	77–80	NA	36
C-80	W	Wide flange section	WTC 1	603	92–95	51	36
C-155	W	Wide flange section	WTC 1	904	83–86	52	36
HH or S-2	W	Wide flange section	WTC 1	605	98–101	53	42

a. C-83 was not positively identified but due to similar size and shape was deemed a core column.
Key: NA, information not available.

Table 3–4. Other built-up box columns and wide flange sections from WTC 1 and WTC 2 with ambiguous stampings and/or markings.

NIST Name	Type	Description	Markings
C-79	RB	Thin rectangular column	101A 81–85–87–92 52
C-101	RB	Thin rectangular column	78A 10 27 50
C-154	RB	Thin rectangular column	825: 107–108 52
C-26	W	Three connected wide flange sections	604/605 107 64 50
C-44	W	Wide flange section	59 S 563
C-45	W	Wide flange section	16 S2 563 F_y 50
C-60	W	Wide flange section	193 S1 69
C-61	W	Wide flange section	150 S 69
C-62	W	Wide flange section	224 (S) <48> F_y 50
M-17	W	Wide flange section	163 (9) 62 F_y 36
M-23	W	Wide flange section	F 2010
M-37	W	Wide flange section	130 (8x–92) <50>
M-38	W	Wide flange section	F_y 42

Note: "x" indicates an unreadable number.

Structural Elements Recovered from the WTC Buildings

Source: NIST.

Figure 3–3. Example of stampings on the interior base of the middle column for each panel.

Chapter 3

Source: NIST.

Figure 3–4. Example of stampings placed on one end of a core column.

a)

Source: NIST.

b)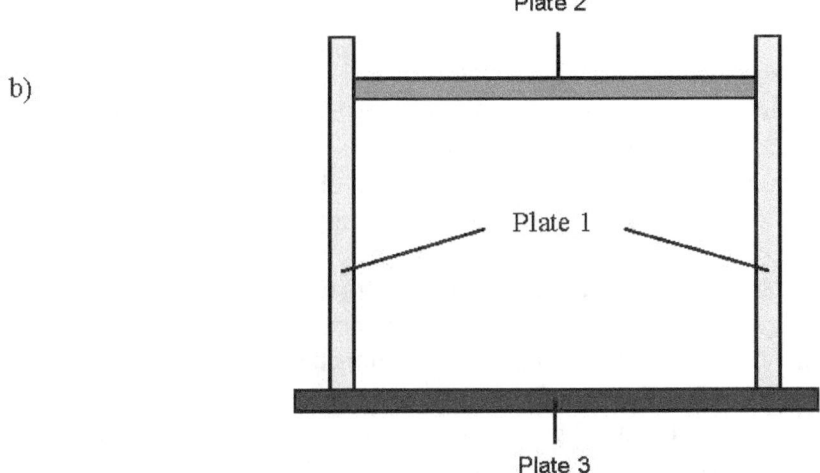

Figure 3–5. (a) Example of stamping placed on flange indicating the column type (120), and (b) schematic indicating the various plates corresponding to Table 3–5.

Table 3–5. Examples of column types with corresponding plate gauges.

Column Type	Plate 1 (in.)	Plate 2 (in.)	Plate 3 (in.)
120	1/4	1/4	1/4
121	5/16	1/4	1/4
122	3/8	1/4	1/4
123	7/16	1/4	1/4
124	1/2	1/4	1/4
125	9/16	1/4	1/4
126	5/8	1/4	1/4
128	3/4	1/4	1/4
129	13/16	5/16	5/16
133	1 1/16	3/8	3/8
149	2 1/16	11/16	11/16
150	2 1/8	3/4	3/4
152	2 1/4	3/4	3/4
334	1 1/8	3/8	3/8
335	1 3/16	7/16	7/16
520	1/4	1/4	1/4
522	3/8	1/4	1/4

Each of the structural elements was additionally stenciled in white or yellow lettering with similar building information. For the exterior panel sections, the stenciling was located on or near the lower spandrel on the interior face. Figure 3–6 (a) shows a typical stenciling found on a perimeter panel, indicating this piece was in WTC 2, with center column line number 300, spanning floors 85 through 87. For the core columns, both stenciling and handwritten codes have been observed on the recovered pieces. Figure 3–6 (b) shows one of these stencilings from a core column located in WTC 1.

Also seen in Fig. 3–6 (a) are two other indicators, 3T and <570>, found on the exterior panel sections. These markings are the estimated piece tonnage (1 ton equals approximately 907 kg) and the erector's derrick division number, respectively. This information was also stamped on some of the core column pieces (see Fig. 3–4). The erector, Karl Koch Erecting Co., Inc., assigned derrick divisions 47 through 70 for WTC 1 and derrick divisions 547 through 570 for WTC 2.[4] Each division was assigned to a specific area of the building and shared a crane with other nearby derrick divisions. Therefore, a single crane may have lifted pieces from derrick divisions 65, 67, and 69. Figure 3–7 shows the derrick division numbers that hoisted the specific columns for both buildings, according to the derrick numbers found on structural elements with positive identification (also shown in Tables 3–2 and 3–3).

---

[4] PONYA (Port of New York Authority). 1967. Communication to Steel Fabricators, Detailers, and Erectors, Shop drawing procedures and marking systems, May 1.

Structural Elements Recovered from the WTC Buildings

a)

b)

Source: NIST.

Figure 3–6. (a) Characteristic stenciling found on the lower portions of the exterior column panels for sample C-14. (b) Characteristic stenciling found on an interior core column for sample B-6152.

Chapter 3

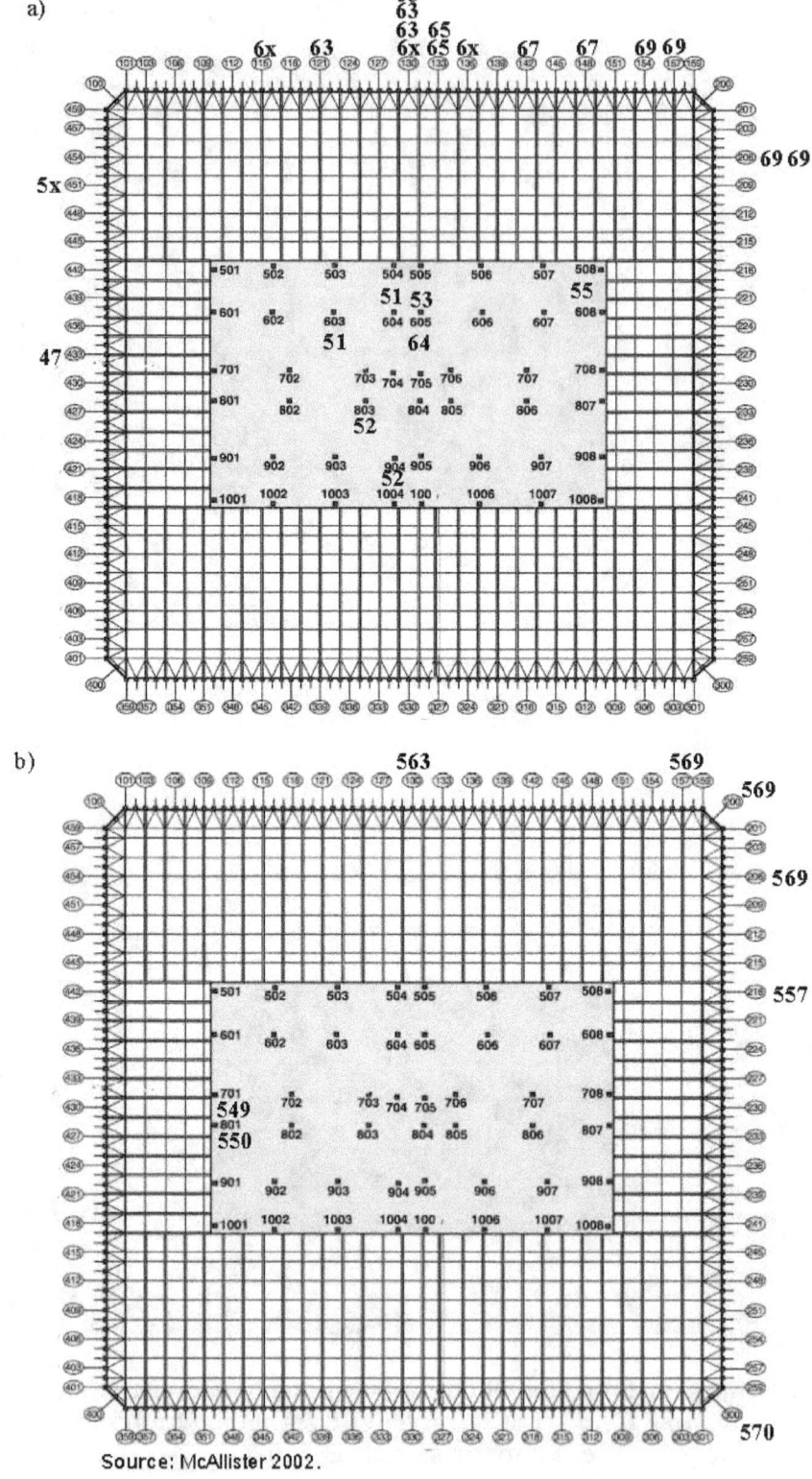

Source: McAllister 2002.

Figure 3–7. Schematic showing derrick divisions that hoisted the specific columns for (a) WTC 1 and (b) WTC 2. The "x" signifies the information that was not readable.

Of the 42 positively identified exterior panels, 25 had specific markings giving all the information needed (building, column, floors) to locate the structural element within the buildings from one or both codes (i.e., stampings or stencils). The flange stampings, which indicated the specified yield strength and column type, were used to confirm the findings (Tables 3–6 and 3–7). The only deviation noted was that 100 ksi steel was substituted for the 85 ksi and 90 ksi grades that were specified. This can be observed in Table 3–6 for samples B-1043, B-1044, C-10, and M-10b. This substitution is consistent with PANYNJ documents of the construction period, indicating that 100 ksi steel was used for all steel specified as 85 ksi or 90 ksi. (See NIST NCSTAR 1-3A, *Contemporaneous Structural Steel Specifications*.)

Seventeen other panels were positively identified using a combination of the stampings, including the specified minimum yield strength (Table 3–8) and column type (Table 3–9), the stenciled derrick division number (Table 3–8), or association to another panel, as follows:

- **ASCE-2**: No information was available signifying the panel identification as only the one outer column was recovered. The flange stampings indicated that the column was of the 356 type with F_y 50 ksi steel. The column had large floor truss seats that fit seat detail 4410 or 4424. Reviewing the building design drawings, seven panels meet the criterion that the leftmost column (when viewed from inside the building looking out) of the panel had the 356 – 50 combination (Table 3–10). Of these, only one panel had the proper floor truss seat identifying the panel as B330: 40-43.

- **C-10**: The stampings indicated that the center column line number was 451 and the panel spanned floors 85 through 88, but the building identification information was obscured by a weld bead. The building can be identified by a derrick division number in the 50 series, which corresponds to WTC 1 (Fig. 3–7). (Note that the flange stampings indicated that the steel used is 100 ksi, while the building design drawings indicated that 85 ksi was specified. As mentioned above, substitution of the specified 85 ksi, as well as the 90 ksi grades, by 100 ksi steel was approved.)

- **C-24**: This piece was readily identifiable as a mechanical or service floor due to the non-uniform width of the columns. Unfortunately, only the upper portion of the panel was recovered, and thus no stampings were found. However, the end connections to these floors were welded in addition to the typical bolting. In doing so, the end plate and a small portion of the column from the panel above this piece remained after the collapse, and the stamping of "B 203 77-78" identifying the panel above this sample was clearly visible.

- **C-55**: The stampings indicated that the center column line number was 209 and the panel spanned floors 94 through 97, however, no building information was observed. By reviewing the flange stampings (Table 3–8), the piece was determined to belong to WTC 1.

- **C-92**: Stenciling on the piece indicated that it was from WTC 2, floors 93 through 96. However, the center column line number was partially obscured, with 13x visible. By reviewing the flange stampings (Tables 3–8 and 3–9), the piece center column line number was determined to be 130.

Chapter 3

Table 3–6. Specified and observed minimum yield strengths for positively identified exterior column panels.[a]

NIST Name	Bldg	Column	Floors	Specified Minimum Yield (ksi)			Stamping Observed		
				Column 1	Column 2	Column 3	Column 1	Column 2	Column 3
ASCE-2	WTC 2	329	40-43	50	50	50	50	NA	NA
B-1024	WTC 2	154	21-24	50	50	50	NA	50	NA
B-1043	WTC 2	406	40-43	85	90	90	100	100	100
B-1044	WTC 2	409	40-43	85	80	85	100	80	100
C-10	WTC 1	451	85-88	85	85	90	100	100	100
C-13 or S-11 and C13a or S-19	WTC 2	200	90-92	100	100	100	100	NA	NA
C-14 or S-18	WTC 2	300	85-87	100	100	100	NA	NA	NA
C-18	WTC 2	230	93-96	55	55	55	55	55	55
C-22	WTC 1	157	93-96	80	75	80	80	NA	80
C-24	WTC 2	203	74-77	100	100	100	NA	NA	NA
C-25	WTC 1	206	89-92	80	80	80	80	NA	NA
C-40	WTC 1	136	98-101	60	60	55	NA	60	55
C-46	WTC 2	157	68-71	80	70	65	80	NA	65
C-48 or S-5	WTC 2	442	91 - 94	65	65	65	NA	65	NA
C-55	WTC 1	209	94-97	70	70	70	NA	70	NA
C-89	WTC 2	215	12 - 15	50	50	55	NA	NA	NA
C-92	WTC 2	130	93 - 96	60	60	60	60	NA	NA
C-93	WTC 1	339	99 - 102	60	60	60	NA	60	NA
CC	WTC 1	124	70-73	50	50	50	NA	50	50
K-1 or K-13	WTC 1	209	97-100	60	60	60	60	60	60
K-2 or K-40	WTC 1	236	92-95	65	65	65	NA	65	NA
M-2	WTC 1	130	96-99	55	55	55	55	55	55
M-10a	WTC 2	209	82-85	85	85	85	NA	NA	NA
M-10b	WTC 2	206	83-86	85	85	85	100	100	NA
M-20	WTC 1	121	99-102	55	55	55	NA	55	55
M-26	WTC 1	130	90-93	50	55	50	NA	55	50
M-27	WTC 1	130	93-96	50	55	55	50	55	NA
M-28	WTC 2	345	98 - 101	70	70	70	NA	NA	NA
M-30	WTC 1	133	94-97	55	55	55	NA	55	55
N-1	WTC 1	218	82-85	70	75	75	70	75	NA
N-7 or M-3	WTC 1	127	97-100	55	55	60	55	55	60
N-8 or M-7	WTC 1	142	97-100	60	60	60	NA	60	NA
N-9 or M-8	WTC 1	154	101-104	55	55	55	55	55	NA
N-10 or M-15	WTC 1	115	89-92	55	55	55	NA	55	55
N-12 or M-13	WTC 1	206	92-95	75	75	75	NA	75	75
N-13 or M-14	WTC 1	130	99-102	55	55	55	NA	NA	NA
N-99 or M-16	WTC 1	148	99-102	65	65	65	65	65	NA
N-101 or M-21	WTC 1	133	100-103	55	55	55	55	55	55
S-1 or EE	WTC 1	433	79-82	70	70	70	NA	70	70
S-9 or C-63	WTC 1	133	97-100	55	55	55	55	55	55
S-10 or C-17	WTC 1	224	92-95	70	70	70	70	70	NA
S-14 or C-20	WTC 2	218	91-94	65	65	70	65	65	70

a. Columns 1, 2, and 3 are viewed left to right as viewed from the inside of the building.
Key: NA, information not available.

Table 3–7. Specified and observed column types for positively identified exterior column panels.[a]

NIST Name	Bldg	Column	Floors	Specified Column Type			Stamping Observed		
				Column 1	Column 2	Column 3	Column 1	Column 2	Column 3
ASCE-2	WTC 2	329	40-43	356	356	356	356	NA	NA
B-1024	WTC 2	154	21-24	149	150	152	149	150	152
B-1043	WTC 2	406	40-43	335	334	334	335	334	334
B-1044	WTC 2	409	40-43	335	335	335	335	335	335
C-10	WTC 1	451	85-88	120	120	120	120	120	120
C-13 or S-11 and C13a or S-19	WTC 2	200	90-92	120	520	120	120	NA	NA
C-14 or S-18	WTC 2	300	85-87	122	522	120	NA	NA	NA
C-18	WTC 2	230	93-96	120	120	120	120	120	120
C-22	WTC 1	157	93-96	120	120	120	120	NA	120
C-24	WTC 2	203	74-77	325	325	325	Bottoms missing		
C-25	WTC 1	206	89-92	120	120	120	120	NA	NA
C-40	WTC 1	136	98-101	121	121	121	NA	121	121
C-46	WTC 2	157	68-71	126	128	129	126	NA	129
C-48 or S-5	WTC 2	442	91 - 94	120	120	120	NA	120	NA
C-55	WTC 2	209	94-97	120	120	120	NA	120	NA
C-89	WTC 2	215	12 - 15	147	145	143	NA	NA	NA
C-92	WTC 2	130	93 - 96	124	123	123	124	NA	NA
C-93	WTC 1	339	99 - 102	121	121	121	NA	121	NA
CC	WTC 1	124	70-73	133	133	133	NA	133	133
K-1 or K-13	WTC 1	209	97-100	120	120	120	120	120	120
K-2 or K-40	WTC 1	236	92-95	120	120	120	NA	120	NA
M-2	WTC 1	130	96-99	122	122	122	122	122	122
M-10a	WTC 2	209	82-85	120	120	120	NA	NA	NA
M-10b	WTC 2	206	83-86	120	120	120	120	120	NA
M-20	WTC 1	121	99-102	120	120	120	NA	120	120
M-26	WTC 1	130	90-93	125	125	125	NA	125	125
M-27	WTC 1	130	93-96	124	123	123	124	123	NA
M-28	WTC 2	345	98 - 101	120	120	120	NA	NA	NA
M-30	WTC 1	133	94-97	123	123	123	NA	123	123
N-1	WTC 1	218	82-85	123	123	123	123	123	NA
N-7 or M-3	WTC 1	127	97-100	121	121	121	121	121	121
N-8 or M-7	WTC 1	142	97-100	121	121	121	NA	121	NA
N-9 or M-8	WTC 1	154	101-104	120	120	120	120	120	NA
N-10 or M-15	WTC 1	115	89-92	125	125	125	NA	125	125
N-12 or M-13	WTC 1	206	92-95	120	120	120	NA	120	120
N-13 or M-14	WTC 1	130	99-102	121	121	120	NA	NA	NA
N-99 or M-16	WTC 1	148	99-102	120	120	120	120	120	NA
N-101 or M-21	WTC 1	133	100-103	120	120	120	120	120	120
S-1 or EE	WTC 1	433	79-82	123	123	123	NA	123	123
S-9 or C-63	WTC 1	133	97-100	122	122	122	122	122	122
S-10 or C-17	WTC 1	224	92-95	120	120	120	120	120	NA
S-14 or C-20	WTC 2	218	91-94	120	120	120	120	120	120

a. Columns 1, 2, and 3 are viewed left to right as viewed from the inside of the building.
Key: NA, information not available.

Table 3–8. Specified minimum yield strengths (ksi) from WTC 1 and WTC 2, along with the observed stampings, used to positively identify some exterior column panels.[a]

NIST Name	Markings	Column line	Floors	Derrick division	If WTC 1 Column 1	If WTC 1 Column 2	If WTC 1 Column 3	If WTC 2 Column 1	If WTC 2 Column 2	If WTC 2 Column 3	Observed Column 1	Observed Column 2	Observed Column 3	Confirmed identification	Full identification
ASCE-2	NA	NA	NA	NA	Column 1 was 50 ksi						50	NA	NA		Inconclusive
C-10	451: 85 - 88	451	85 - 88	5x	90	85	80	80	80	80	100	100	100	WTC 1	A451: 85-88
C-55	209: 94 - 97	209	94 - 97	NA	70	70	60	60	60	60	NA	70	NA	WTC 1	A209: 94-97
C-92	B13x 93-96	130	93 - 96	NA	"B" indicates WTC 2	60	60	60	60	60	60	60	NA	130	B130: 93 - 96
		139	93 - 96	NA			65	60	60						
C-93	339: 99 - 102	339	99 - 102	NA	60	60	60	65	65	60	NA	60	NA	WTC 1	A339: 99 - 102
CC	124: 70 - 73	124	70 - 73	NA	50	50	55	55	55	55	NA	50	50	WTC 1	A124: 70-73
K-1	209: 97 - 100	290	97 - 100	NA	60	60	60	55	55	60	60	60	60	WTC 1	A209: 97-100
K-2	236: 92 - 95	236	92 - 95	NA	65	65	65	60	60	60	NA	65	NA	WTC 1	A236: 92-95
M-2	x - 9x <63>		90's	63	3 columns of column type 122						55	55	55		Inconclusive
M-30	x33: 94 - 97	133	94 - 97	65	55	55	60	60	60	60	NA	55	55	WTC 1, 133	A133: 94-97
		233	94 - 97		60	60	55	55	55	55					
		333	94 - 97		55	55	60	60	60	55					
		433	94 - 97		65	65	55	55	55	50					
N-1	2x8: 82 - 85	218	82 - 85	5x	70	75	60	60	60	65	70	75	NA	WTC 1, 218	A218: 82-85
		248	82 - 85		Column line 248 spans either floors 81 - 84 or 84 - 87										
N-7	127: 97 - 100	127	97 - 100	NA	60	55	60	65	65	60	60	55	55	WTC 1	A127: 97-100
N-12	x06: 92 - 95	106	92 - 95	69	65	65	65	65	65	70	NA	75	75	WTC 1, 206	A206: 92-95
		206	92 - 95		75	75	65	65	65	65					
		306	92 - 95		65	65	65	65	65	70					
		406	92 - 95		70	70	70	70	70	70					
S-10 or C-17	224: 92 - 95	224	92 - 95	NA	70	70	60	60	60	60	70	70	NA	WTC 1	A224: 92-95

a. Columns 1, 2, and 3 are left to right viewed from inside the building.
Key: NA, information not available.

Table 3-9. Specified column types of exterior panels from WTC 1 and WTC 2, along with the observed stampings, used to positively identify some exterior column panels.[a]

NIST Name	Markings	Column line	Floors	If WTC 1 Column 1	If WTC 1 Column 2	If WTC 1 Column 3	If WTC 2 Column 1	If WTC 2 Column 2	If WTC 2 Column 3	Observed Column 1	Observed Column 2	Observed Column 3	Confirmed identification
ASCE-2	NA	NA	NA	Column 1 was of type 356						356	NA	NA	Inconclusive
C-10	451: 85 - 88	451	85 - 88	120	120	120	120	120	120	120	120	120	Inconclusive
C-55	209: 94 - 97	209	94 - 97	120	120	120	120	120	120	NA	120	NA	Inconclusive
C-92	B13x: 93-96	130 / 139	93 - 96 / 93 - 96	"B" indicates WTC 2			124 / 123	123 / 124	123 / 124	124	NA	NA	130
C-93	339: 99 - 102	339	99 - 102	121	121	121	121	121	121	NA	121	NA	Inconclusive
CC	124: 70 - 73	124	70 - 73	133	133	133	133	133	133	133	133	NA	Inconclusive
K-1	209: 97 - 100	290	97 - 100	120	120	120	120	120	120	120	120	120	Inconclusive
K-2	236: 92 - 95	236	92 - 95	120	120	120	120	120	120	NA	120	NA	Inconclusive
M-2	x - 9x <63>		90's				3 columns of having 55 ksi			122	122	122	Inconclusive
M-30	x33: 94 - 97	133 / 233 / 333 / 433	94 - 97 / 94 - 97 / 94 - 97 / 94 - 97	123 / 120 / 123 / 120	123 / 120 / 123 / 120	123 / 120 / 123 / 120	123 / 120 / 123 / 120	123 / 120 / 123 / 120	123 / 120 / 123 / 120	123	123	NA	233 and 433 eliminated
N-1	2x8: 82 - 85	218 / 248	82 - 85 / 82 - 85	123	123	123	123	123	123	NA	123	123	218
				Column line 248 spans either floors 81 - 84 or 84 - 87									
N-7	127: 97 - 100	127	97 - 100	121	121	121	121	121	121	121	121	121	Inconclusive
N-12	x06: 92 - 95	106 / 206 / 306 / 406	92 - 95 / 92 - 95 / 92 - 95 / 92 - 95	122 / 120 / 122 / 120	122 / 120 / 122 / 120	122 / 120 / 122 / 120	122 / 120 / 122 / 120	122 / 120 / 122 / 120	122 / 120 / 122 / 120	120	120	NA	106 and 306 eliminated
S-10 or C-17	224: 92 - 95	224	92 - 95	120	120	120	120	120	120	NA	120	120	Inconclusive

a. Columns 1, 2, and 3 are left to right viewed from inside the building.
Key: NA, information not available.

Chapter 3

Table 3–10. Information used to determine the identification of exterior panel ASCE-2.

1) 7 panels from towers with C1[a] having type 356 column and 50 Fy

	PANEL NUMBER			COLUMN 1						COLUMN 2					COLUMN 3				
	Center Col #	Story @ Splice		PANEL TYPE	Type	FY1 (ksi)	FY2 (ksi)	Splice Detail		Type	FY1 (ksi)	FY2 (ksi)	Splice Detail		Type	FY1 (ksi)	FY2 (ksi)	Splice Detail	
Bldg		Upper	Lower					CSU	CSL				CSU	CSL				CSU	CSL
	(1)	(2)	(3)	(4)	(5)	(6)	(7)	(8)	(9)	(5)	(6)	(7)	(8)	(9)	(5)	(6)	(7)	(8)	(9)
WTC 1	133	43	40	401	356	50	50	103	103	356	46	46	103	103	356	46	46	103	103
WTC 1	142	43	40	401	356	50	50	102	103	355	50	50	102	103	355	50	50	102	103
WTC 2	330	43	40	400	356	50	50	102	103	356	50	50	102	103	356	50	50	103	103
WTC 2	333	43	40	400	356	50	50	103	103	356	50	50	103	103	356	50	50	103	103
WTC 2	336	43	40	400	356	50	50	103	103	356	50	50	103	103	356	50	50	103	103
WTC 2	339	43	40	400	356	50	50	103	103	356	50	50	103	103	356	50	50	102	103
WTC 2	342	43	40	400	356	50	50	102	103	355	55	55	102	103	355	55	55	102	103

2) Only 1 panel has a type 4424 truss seat on C1[a]

	PANEL NUMBER				SPANDREL PLATE				SPANDREL 1					SEAT DETAIL		
	Center Col #	Story @ Splice		FLOOR	T4 (in)	FY4 (ksi)	SCL	SCR	WELDMENTS					COL 1	COL 2	COL 3
		Upper	Lower						NO. 5 t (in)	L (in)	NO. 6 (in)	NO. 7 (in)				
	(1)	(2)	(3)	(10)	(11)	(12)	(13)	(14)	(15)	(16)	(17)	(18)		(19)	(19)	(19)
WTC 1	133	43	40	43	0.9375	42	404	404	0.875	16	0.5	0		6220	4424	6120
	142	43	40	43	0.9375	42	405	405	1.0625	16	0.5	0		4324	6220	4324
WTC 2	330	43	40	43	0.9375	36	403	404	0.875	16	0.5	0		4424	6220	4424
	333	43	40	43	0.9375	36	404	404	0.9375	16	0.5	0		6220	4424	6120
	336	43	40	43	0.9375	36	404	404	1	16	0.5	0		4324	6120	4324
	339	43	40	43	0.9375	36	404	404	1.0625	16	0.5	0		6220	4324	6220
	342	43	40	43	0.9375	36	404	405	1.0625	16	0.5	0		4324	6220	4324

a. Columns 1, 2, and 3 are left to right viewed from inside the building.

- **C-93**: The stampings indicated that the center column line number was 339 and the panel spanned floors 99 through 102; however, no building information was observed. By reviewing the flange stampings (Table 3–8), the piece was determined to belong to WTC 1.

- **CC**: The stampings indicated that the center column line number was 124 and the panel spanned floors 70 through 73; however, no building information was observed. By reviewing the flange stampings (Table 3–8), the piece was determined to belong to WTC 1.

- **K-1**: The stampings indicated that the center column line number was 209 and the panel spanned floors 97 through 100; however, no building information was observed. By reviewing the flange stampings (Table 3–8), the piece was determined to belong to WTC 1.

- **K-2**: The stampings indicated that the center column line number was 236 and the panel spanned floors 92 through 95; however, no building information was observed. By reviewing the flange stampings (Table 3–8), the piece was determined to belong to WTC 1.

- **M-2**: No information was available from the stampings at the base of the middle column, and very little information was recovered from the stenciling on the spandrel. A derrick division number of <63> was observed, placing the element in WTC 1 (Table 3–8). The only other information was – 9, indicating that some portion of the panel was located in the 90s-floor-level range. The flange stampings from the recovered piece specified that all three columns were of the 122 type, with F_y 55 ksi steel. In addition, columns 1 and 3 had floor truss seats, while column 2 had gusset plates for the diagonal bracing straps. Reviewing the building design drawings, it was found that five panels meet the 122 column type, with 55 ksi steel in the 90s range (Table 3–11). Of these, only two panels had columns 1 and 3 with floor truss seats (130: 96 through 99 and 330: 96 through 99). As shown in Fig. 3–7, the derrick division of <63> identifies the panel as 130: 96 through 99.

- **M-10a**: The sample was identified solely by association to another panel (bolted spandrel connection). The sample M-10 retrieved by SEAoNY was actually composed of pieces from two different exterior column panels (Fig. 3–8). Therefore, with the positive identification of M-10b via the stampings and stencils, M-10a's connection to it allowed its identification as WTC 2,209: 82 through 85.

- **M-28**: The stampings indicated that the center column line number was 345 and the panel was located in WTC 2. However, the markings of the floors spanned were partially obscured; 9x–1xx. By reviewing the building design drawings, the only panel that could fit spanned floors 98 through 101.

- **M-30**: The stampings found were x33 94–97, where the "x" signifies missing information due to a weld bead running across this area. Thus, the building and exact center column line numbers were unknown. However, a derrick division number of <65> was visible on the interior spandrel. From this information, as well as the specified minimum yield strength (Table 3–8) and column type (Table 3–9), M-30 was determined to belong to WTC 1, with a center column line number of 133.

Chapter 3

Table 3–11. Information used to determine the identification of exterior panel M-2.

1) 5 panels in WTC 1, floors 90-99 that meet the criterion of 3 columns[a] with column type 122 and 55 ksi

PANEL NUMBER	Story @ Splice		SPANDREL 1				SPANDREL 2				SPANDREL 3			
				SEAT DETAIL				SEAT DETAIL				SEAT DETAIL		
Center Col #	Lower	Upper	FLOOR	COL 1	COL 2	COL 3	FLOOR	COL 1	COL 2	COL 3	FLOOR	COL 1	COL 2	COL 3
127	94	97	97	5110	1411	5210	96	5110	1411	5210	95	5110	1411	5210
130	96	99	99	1411	5210	1411	98	1411	5210	1411	97	1411	5210	1411
330	96	99	99	1411	5210	1411	98	1411	5210	1411	97	1411	5210	1411
133	97	100	100	5210	1411	5110	99	5210	1411	5110	98	5210	1411	5110
333	97	100	100	5210	1411	5110	99	5210	1411	5110	98	5210	1411	5110

2) Only two panels that meet the additional criterion of columns 1 and 3 having truss seat attachments and column 2 having gusset plate attachments
 - Seat detail 5110 and 5120 are gusset plates for diagonal bracing straps
 - Seat detail 1411 are truss seat attachments

PANEL NUMBER	Story @ Splice		SPANDREL 1				SPANDREL 2				SPANDREL 3			
				SEAT DETAIL				SEAT DETAIL				SEAT DETAIL		
Center Col #	Lower	Upper	FLOOR	COL 1	COL 2	COL 3	FLOOR	COL 1	COL 2	COL 3	FLOOR	COL 1	COL 2	COL 3
130	96	99	99	1411	5210	1411	98	1411	5210	1411	97	1411	5210	1411
330	96	99	99	1411	5210	1411	98	1411	5210	1411	97	1411	5210	1411

3) Derrick Division suggests that panel came from North face of WTC 1, i.e., panel in the 100-series

PANEL NUMBER	Story @ Splice		SPANDREL 1				SPANDREL 2				SPANDREL 3			
				SEAT DETAIL				SEAT DETAIL				SEAT DETAIL		
Center Col #	Lower	Upper	FLOOR	COL 1	COL 2	COL 3	FLOOR	COL 1	COL 2	COL 3	FLOOR	COL 1	COL 2	COL 3
130	96	99	99	1411	5210	1411	98	1411	5210	1411	97	1411	5210	1411

a. Columns 1, 2, and 3 are left to right viewed from inside the building.

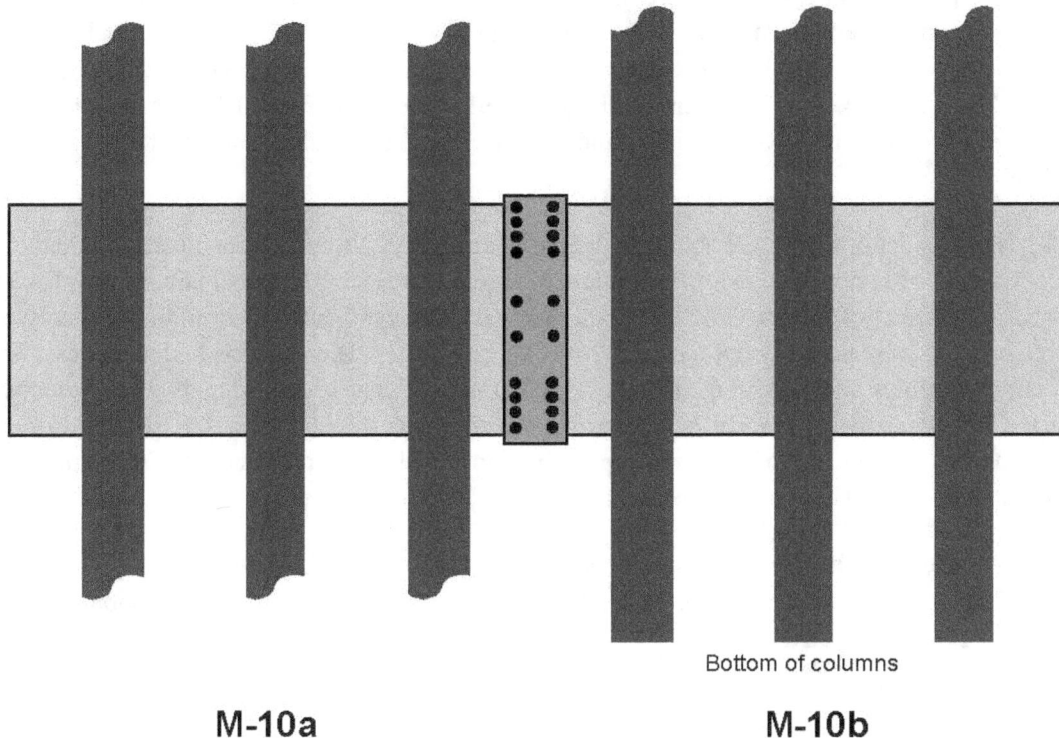

Figure 3–8. Schematic showing the sample M-10 as two separate exterior column panels, M-10a and M-10b.

- <u>N-1</u>: The stampings indicated that the columns spanned floors 82 through 85; however, no building information was observed and a weld bead ran through the middle of the center column line number yielding only "2x8." By reviewing the building plans, only column line 218 spanned the floors specified and the flange stampings (Tables 3–8 and 3–9) indicated that the piece belonged to WTC 1.

- <u>N-7</u>: The stampings indicated that the center column line number was 127 and spanned floors 97 through 100, however, no building information was observed. By reviewing the flange stampings (Table 3–8), the piece belonged to WTC 1.

- <u>N-12</u>: The stampings found were x06 92–95 where the "x" signifies missing information due to a weld bead running across this area. Thus, the building and exact center column line numbers were unknown. However, a derrick division number of <69> was visible on the interior spandrel. From this information, as well as the specified minimum yield strength (Table 3–8) and column type (Table 3–9), it was determined that N-12 belonged to WTC 1, with a center column line number of 206.

- <u>S-10 or C-17</u>: The stampings indicated that the center column line number was 224 and the panel spanned floors 92 through 95; however, no building information was observed. By reviewing the flange stampings (Table 3–8), the piece was determined to belong to WTC 1.

In addition to the overall images taken for record-keeping purposes, the exterior column panels were mapped to indicate how much of the panel was recovered after the collapse. Figure 3–9 displays schematics of typical exterior panels recovered, and Figs. 3–10 and 3–11 show these maps, with the recovered portion indicated, for the identified samples from WTC 1 and WTC 2, respectively. Special note should be given to the fact that these diagrams are drawn as if viewed from the outside of the building. B-1043, B-1044, and C-24 were samples located at the mechanical floors of the building. C-13 and C-13a (pieces of the same exterior panel) and C-14 were exterior wall panels located at the corner of the building.

For the 12 samples identified as core column material (Table 3–3), all but 2 were clearly marked. Figures 3–12 and 3–13 show the portion of column recovered for each individual column from WTC 1 and WTC 2, respectively. Sample C-30 had markings that clearly indicated the building and column; however, the floors were partially obscured: "x04–10x." As the 24 ft section has both connector ends, it spanned only two floors and fit with the floor levels of 104–106. The second sample was C-88b, which did not have any stampings or markings, but was welded to C-88a (identified by stampings). A final sample, C-83, was also found among this group. While no markings were found on the sample, it was recorded as a core column due to its shape, which was very similar to C-90.

There were 13 other wide flange sections that had stampings and/or markings that did not correspond to the code as discussed above (Table 3–4). Instead, there were typically three distinct grouping of numbers and/or letters. Two examples are:

        Sample C-44:   "59   S   563"

        Sample M-17:   "163   9   62"

One piece, C-26 (Fig. 3–14), was distinct among this group in that it was composed of three wide flange sections bolted together at two different angles. The markings on the piece indicated that the wide flanges were 50 ksi steel and came from the 107th floor of WTC 1. Reviewing design drawings, it was found that this piece was a component of the framed floor outside of the core. The as-built location of the other 12 components could not be determined nor confirmed that they were part of the structural steel used in the WTC towers.

Floor trusses were also recovered; however, attempts to identify their specific as-built locations within the buildings were not successful. No stampings were found. Of the 23 pieces held by NIST, 8 are of significant size but are badly tangled and twisted as a result of the collapse and subsequent handling of the material. The remaining pieces consist of shorter sections of chord and rod material in addition to welded sections that connected the trusses to the floor seats.

At present, there are seven samples from WTC 5, all in the GZ-series (see Table A–10). These are coupons that were removed at the WTC site and held by Gilsanz Murray Steficek, LLP. They were subsequently sent to NIST once the Investigation officially began.

No structural elements have been positively identified from WTC 7. However, the columns were fabricated from conventional 36 ksi, 42 ksi, and 50 ksi steel that complied with ASTM International specifications.

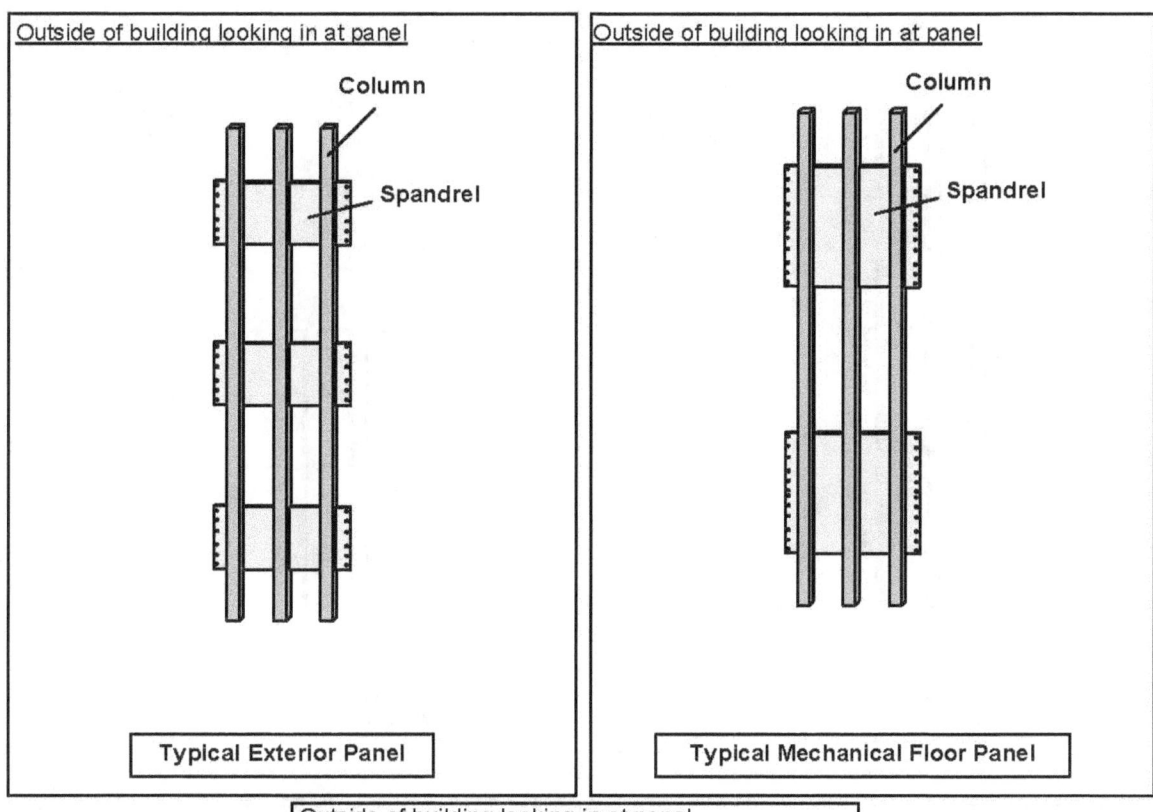
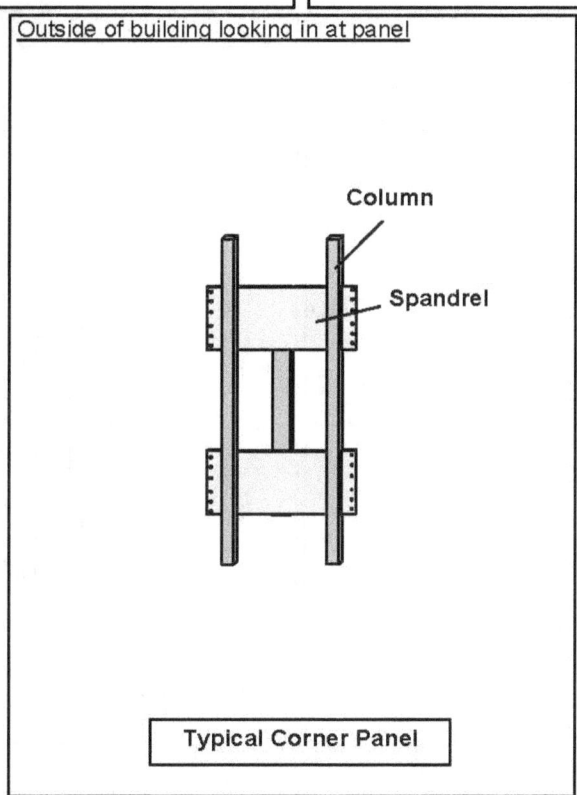

Figure 3–9. Schematics displaying the various types of exterior column panels.

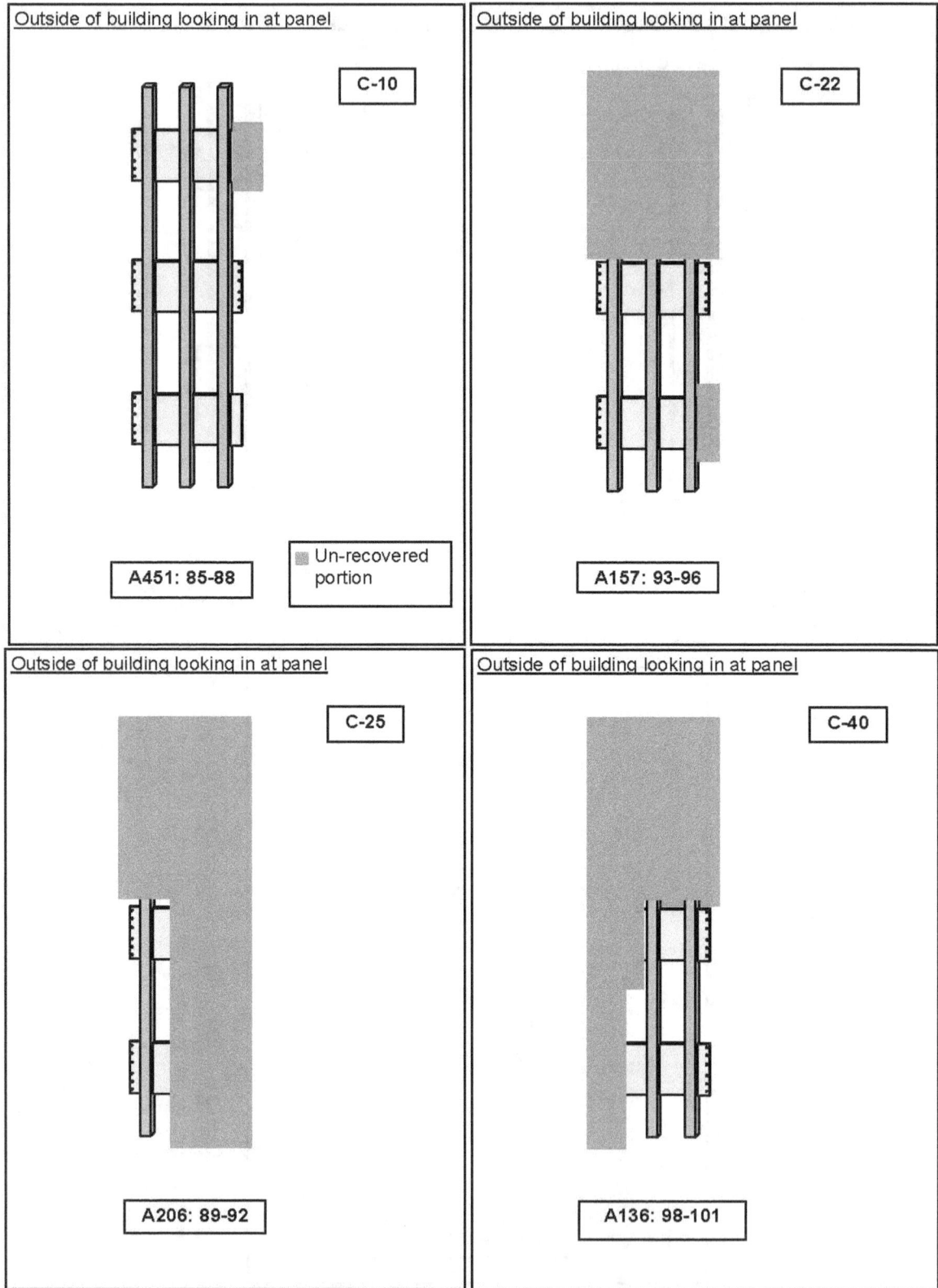

Figure 3–10. Exterior column panel maps indicating the portion of the specific exterior column panel section recovered from WTC 1.

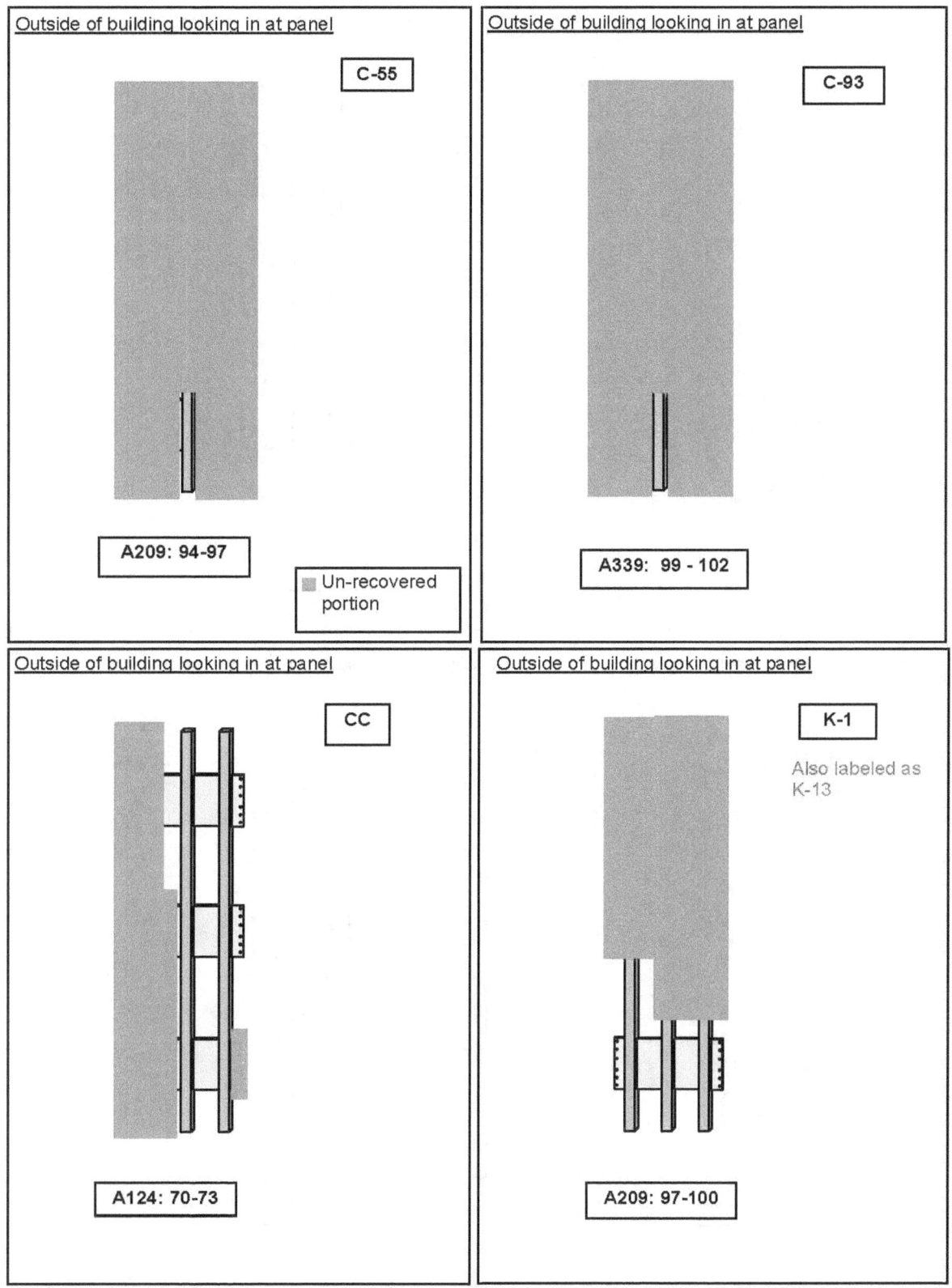

Figure 3–10. Exterior column panel maps indicating the portion of the specific exterior column panel section recovered from WTC 1 (continued).

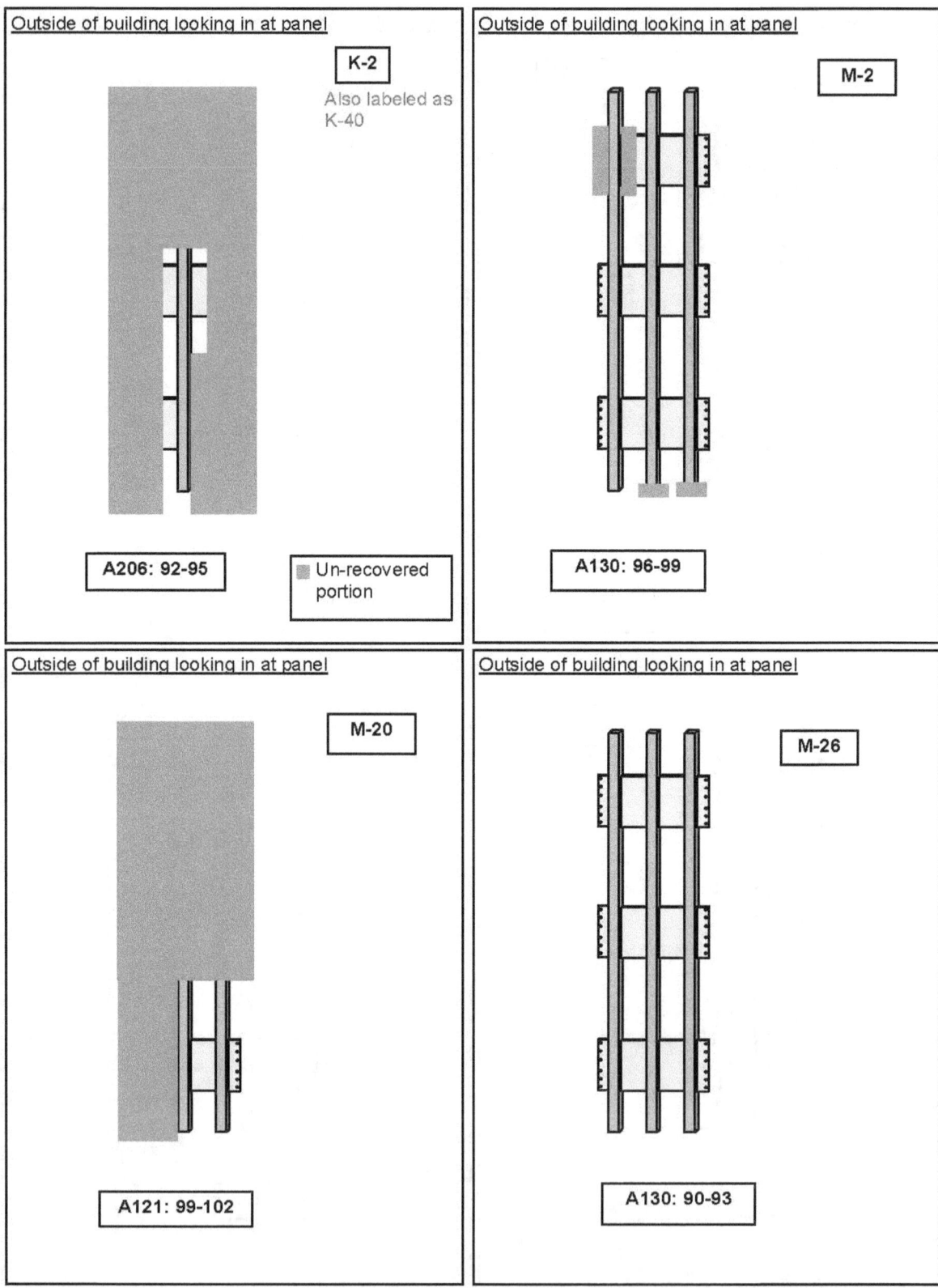

Figure 3–10. Exterior column panel maps indicating the portion of the specific exterior column panel section recovered from WTC 1 (continued).

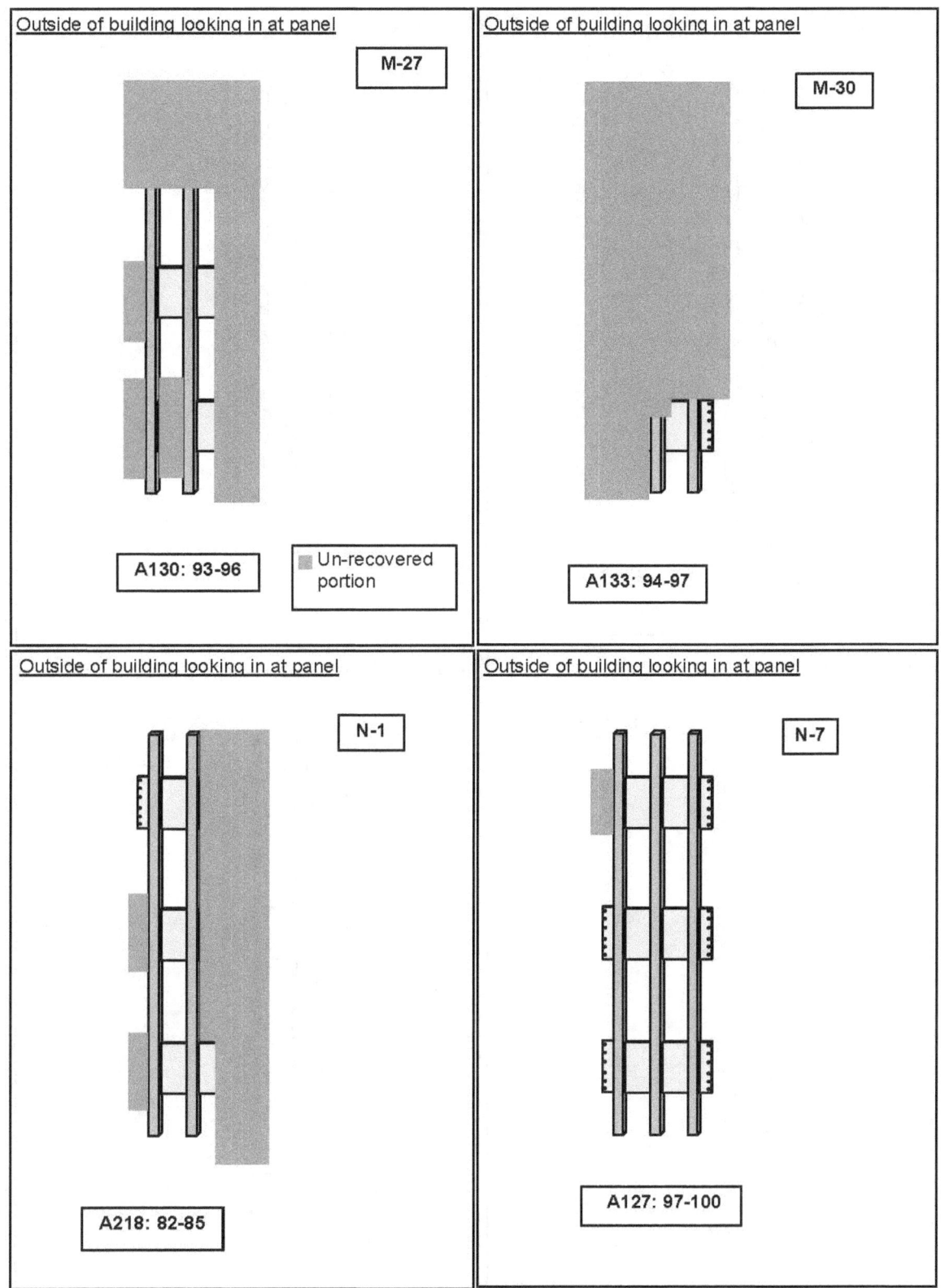

Figure 3–10. Exterior column panel maps indicating the portion of the specific exterior column panel section recovered from WTC 1 (continued).

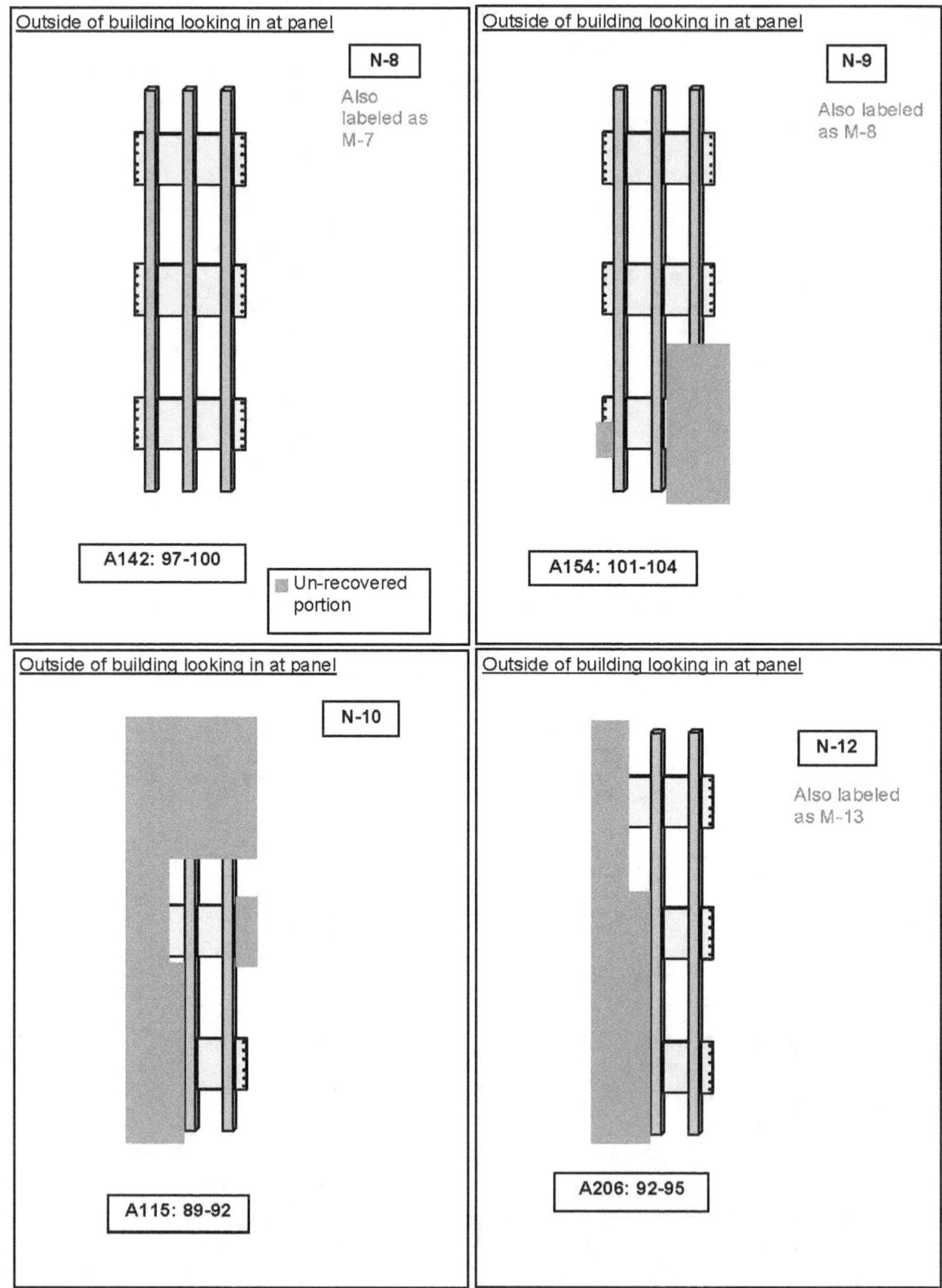

Figure 3–10. Exterior column panel maps indicating the portion of the specific exterior column panel section recovered from WTC 1 (continued).

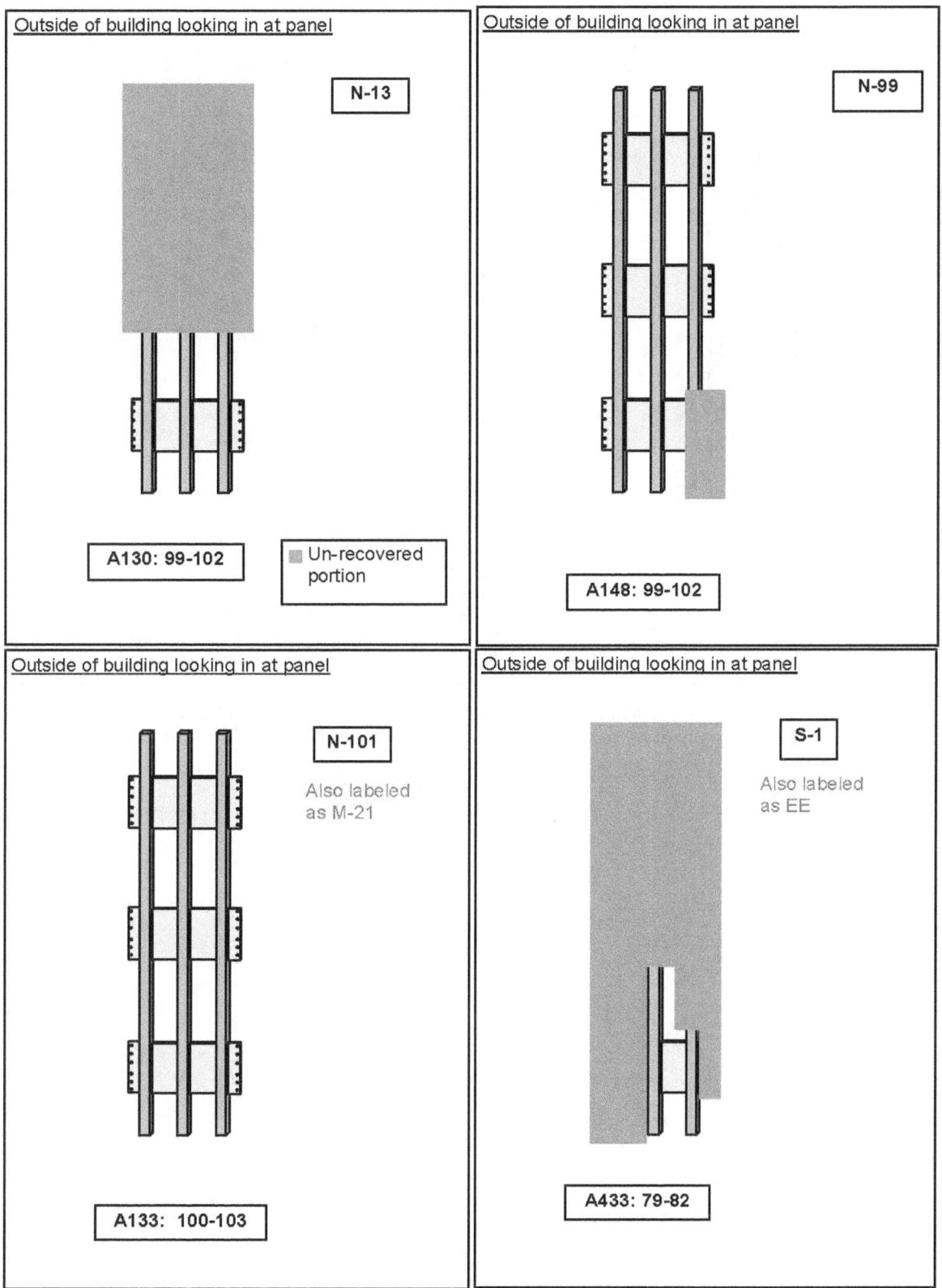

Figure 3–10. Exterior column panel maps indicating the portion of the specific exterior column panel section recovered from WTC 1 (continued).

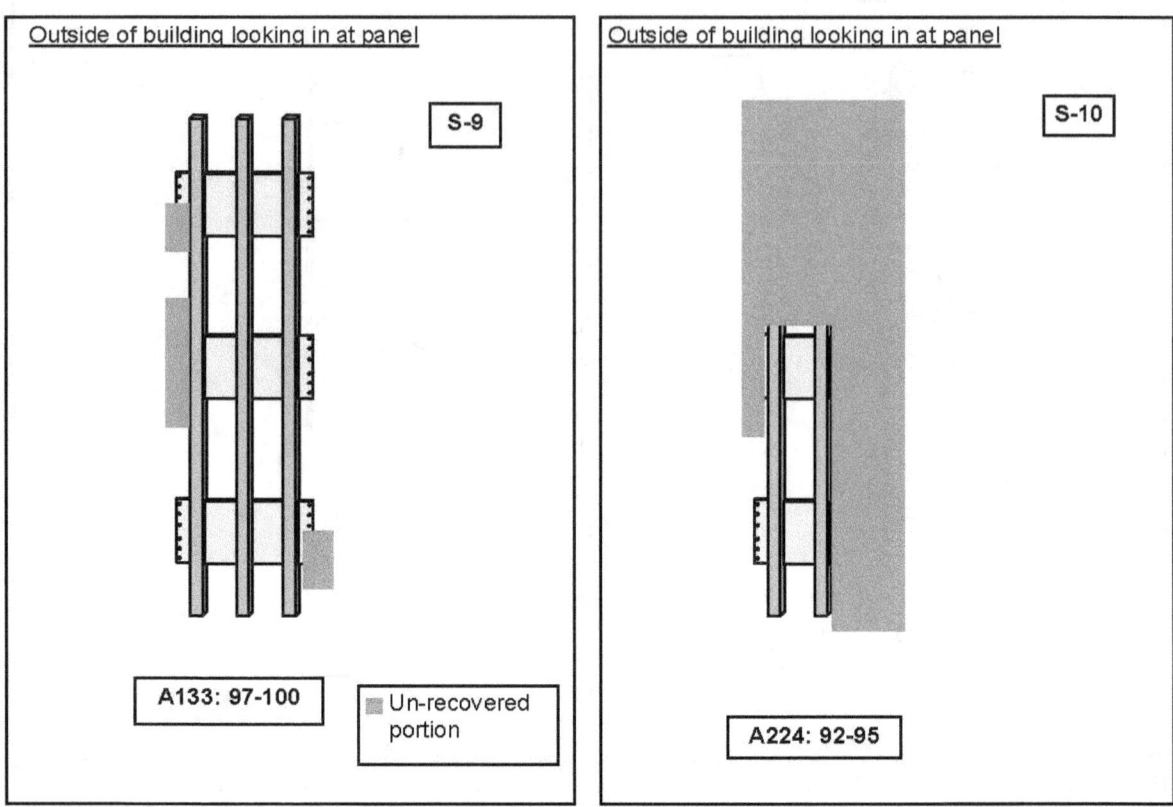

Figure 3–10. Exterior column panel maps indicating the portion of the specific exterior column panel section recovered from WTC 1 (continued).

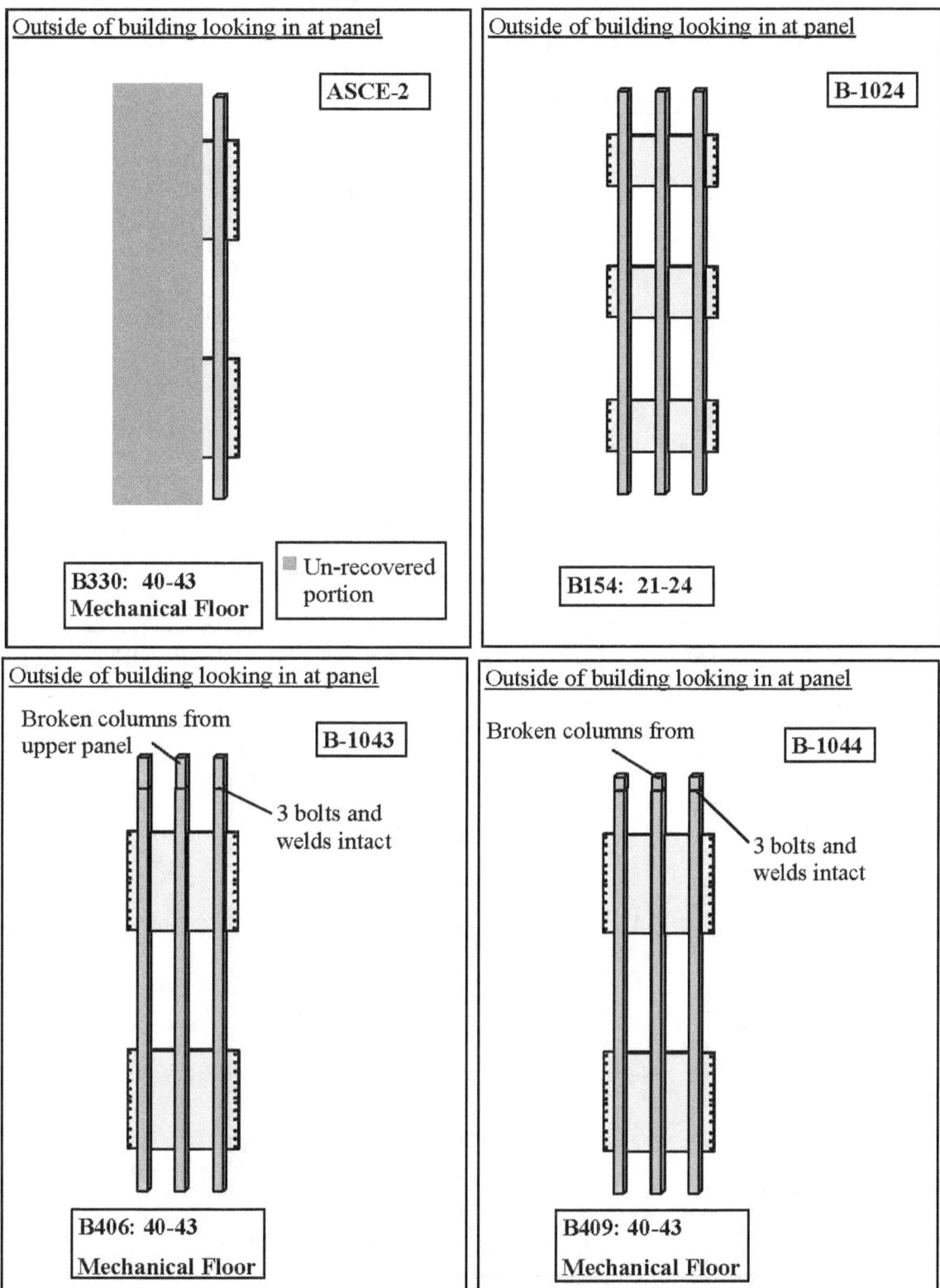

Figure 3–11. Exterior column panel maps indicating the portion of the specific exterior column panel section recovered from WTC 2.

Chapter 3

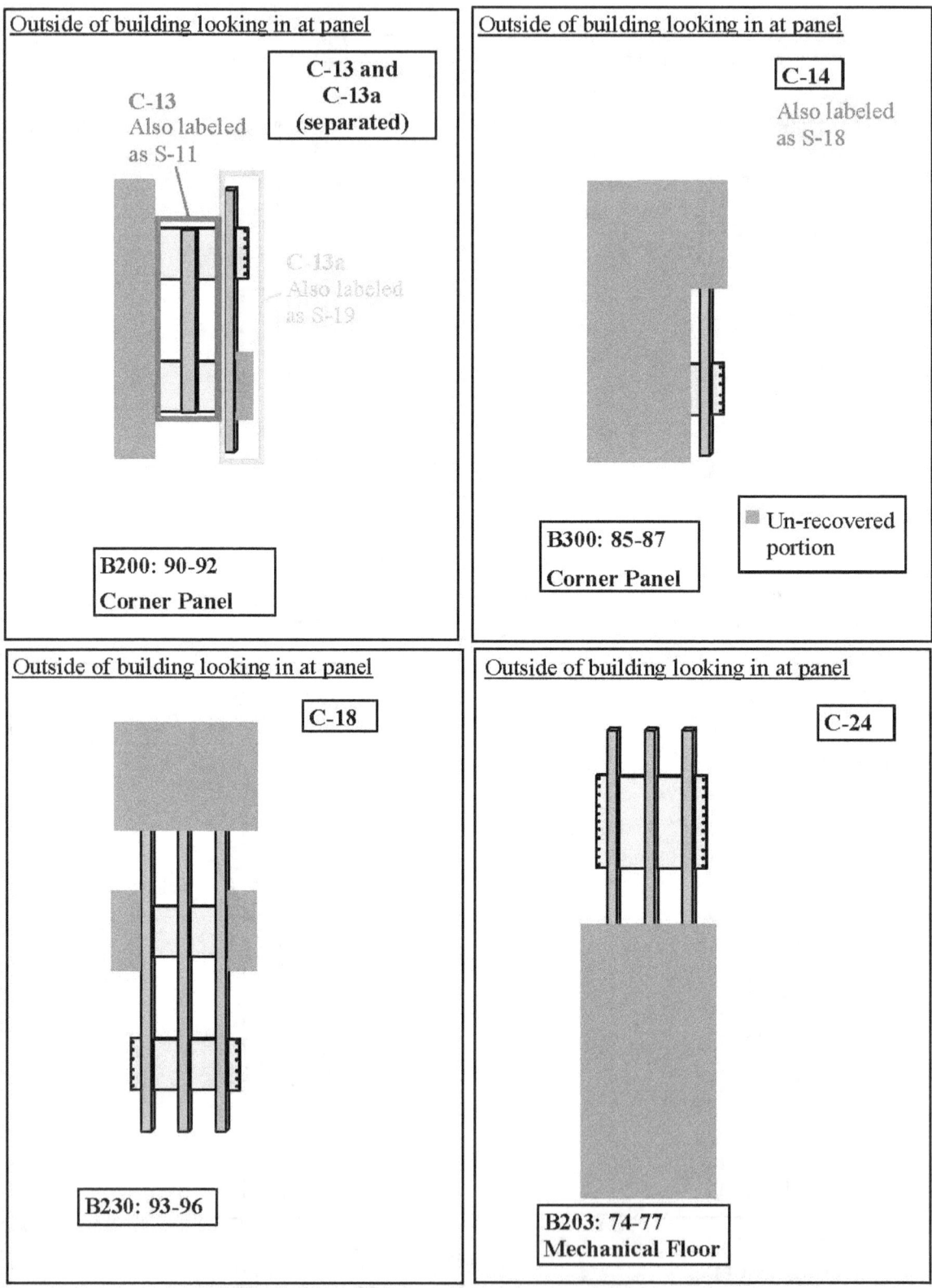

Figure 3–11. Exterior column panel maps indicating the portion of the specific exterior column panel section recovered from WTC 2 (continued).

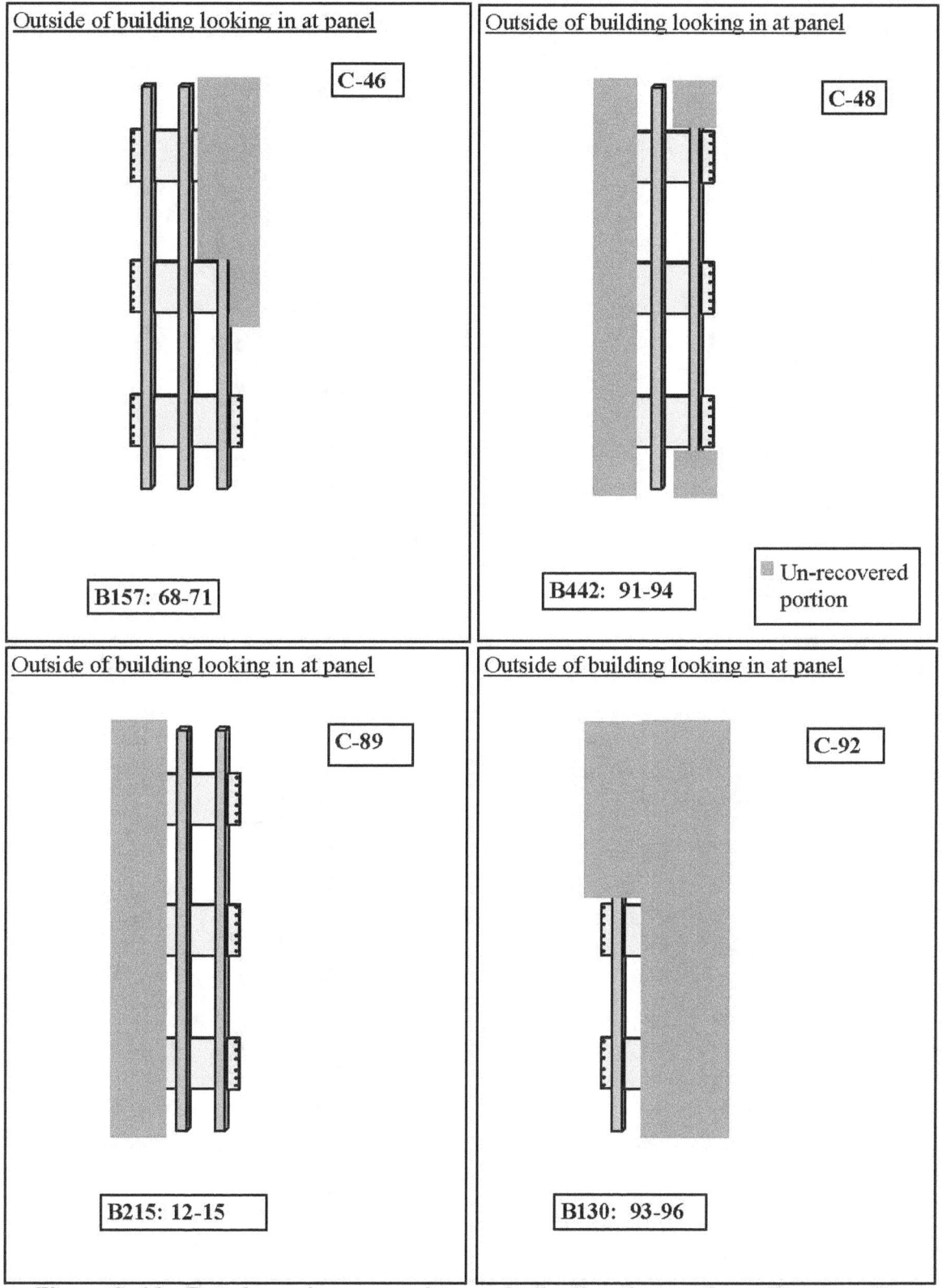

Figure 3–11. Exterior column panel maps indicating the portion of the specific exterior column panel section recovered from WTC 2 (continued).

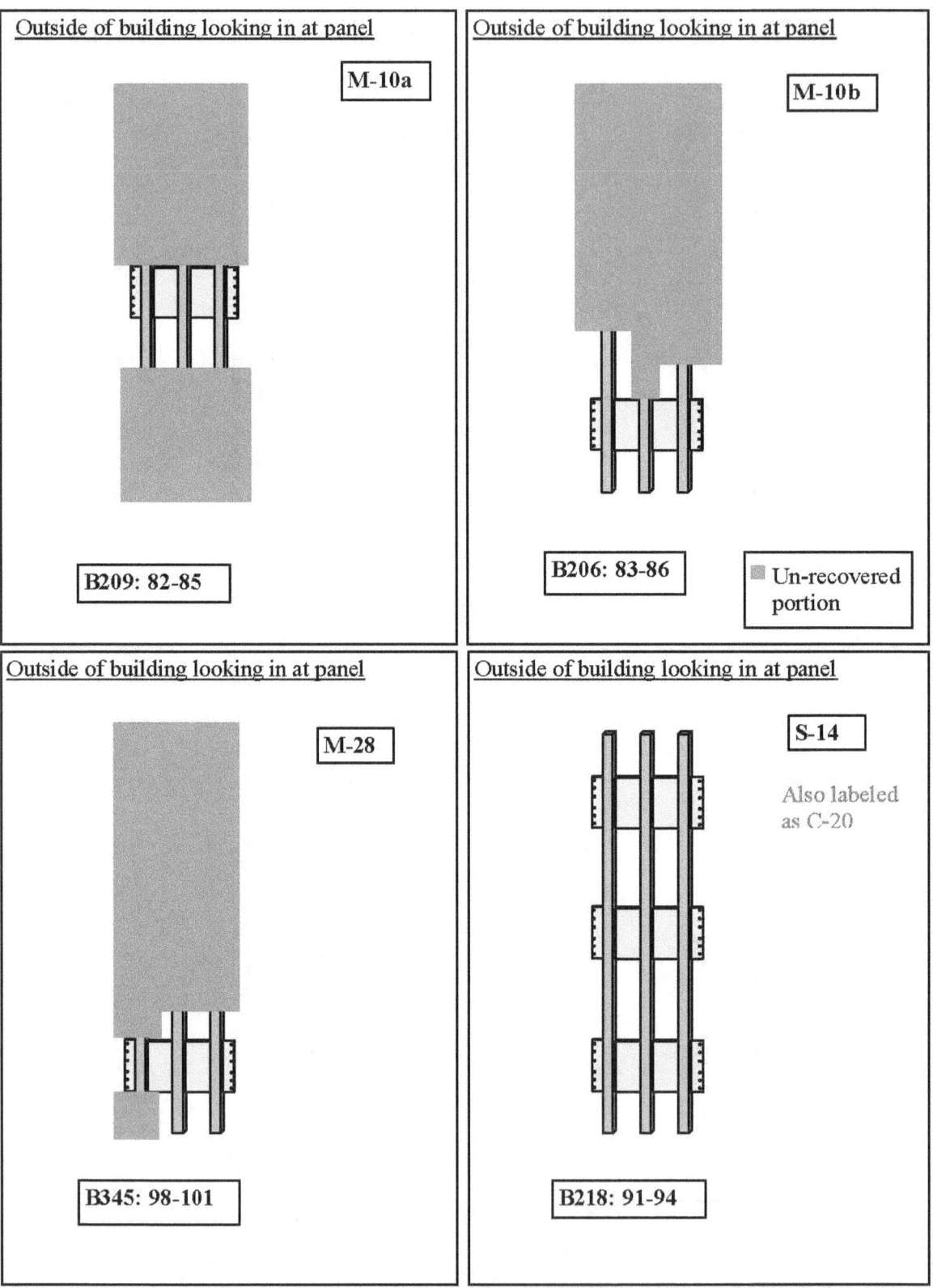

Figure 3–11. Exterior column panel maps indicating the portion of the specific exterior column panel section recovered from WTC 2 (continued).

Source: NIST.

Figure 3–12. Core columns recovered from WTC 1. a) B-1011 (508A: 51–54), lower 2 ft to 3 ft of built-up box column, b) B-6152-1 (803A: 15–18), lower 3 ft of built-up box column.

Source: NIST.

Figure 3–12. Core columns recovered from WTC 1. c) B-6152-2 (504A: 33–36), lower 2 ft of built-up box column, d) C-65 (904A: 86–89), lower 24 ft of wide flange section (continued).

Structural Elements Recovered from the WTC Buildings

Source: NIST.

Figure 3–12. Core columns recovered from WTC 1. e) C-71 (904A: 77–80), lower 16 ft of wide flange section, f) C-80 (603A: 92–95), lower 13 ft of wide flange section (continued).

Source: NIST.

Figure 3–12. Core columns recovered from WTC 1. g) C-155 (904A: 83–86), lower 33 ft of wide flange section, and h) HH (605A: 98–101), lower 16 ft of wide flange section (continued).

Source: NIST.

Figure 3–13. Core columns recovered from WTC 2. a) C-88a (801B: 80–83), lower 16 ft of built-up box column and C-88b (801B: 77–80), upper 8 ft of built-up box column.

Chapter 3

Source: NIST.

Figure 3–13. Core columns recovered from WTC 2. b) C-90 (701B: 12–15), entire length of built-up box column, and c) C-30 (1008B: 104–106) entire length of wide flange section (continued).

Source: NIST.

Figure 3–14. Structural element composed of three wide flange sections bolted together. The component was found to be from the framed floor area outside of the core on the 107th floor of WTC 1 (sample was C-26).

Chapter 4
STRUCTURAL STEEL ELEMENTS OF SPECIAL IMPORTANCE

Of the 41 exterior column panels and 12 core columns positively identified, many were considered especially important to this Investigation led by the National Institute of Standards and Technology (NIST). Two major categories of steel are considered to be of special value:

- Samples located in or around the floors impacted by the airplane

- Samples that can represent 1 of 12 grades of steel specified for the exterior columns, 1 of 4 grades of steel specified for the core columns, and 1 of the 2 grades of steel for the floor trusses

4.1 SAMPLES LOCATED IN OR AROUND THE FLOORS IMPACTED BY THE AIRPLANE

Interpretation of the photographic evidence revealed that damage to World Trade Center (WTC) 1 due to aircraft impact occurred from floor 94 to floor 99 and was bounded by columns 111 through 152. For WTC 2, the impact area was lower with damage found from floor 77 to floor 85. While the damage appears to be bordered by column lines 411 and 440, columns closer to the southeast corner of the building may also have been affected. However, few images were obtained where smoke is not obscuring this portion of the south face of WTC 2 to complete the analysis. From this information, NIST was able to determine which perimeter panels and core columns could be used to comment on damage and possible failure mechanisms in this area. Figure 4–1 shows the sample overlay of the exterior panels in NIST's possession in and around the impact zone of WTC 1. Samples C-80 and HH, both core columns, were also identified as residing near the impact zone. The recovered portion of each column is approximately represented in this image. Unfortunately, there were no similar corresponding exterior panels for WTC 2, but two core columns were recovered (Fig. 4–2).

Chapter 4

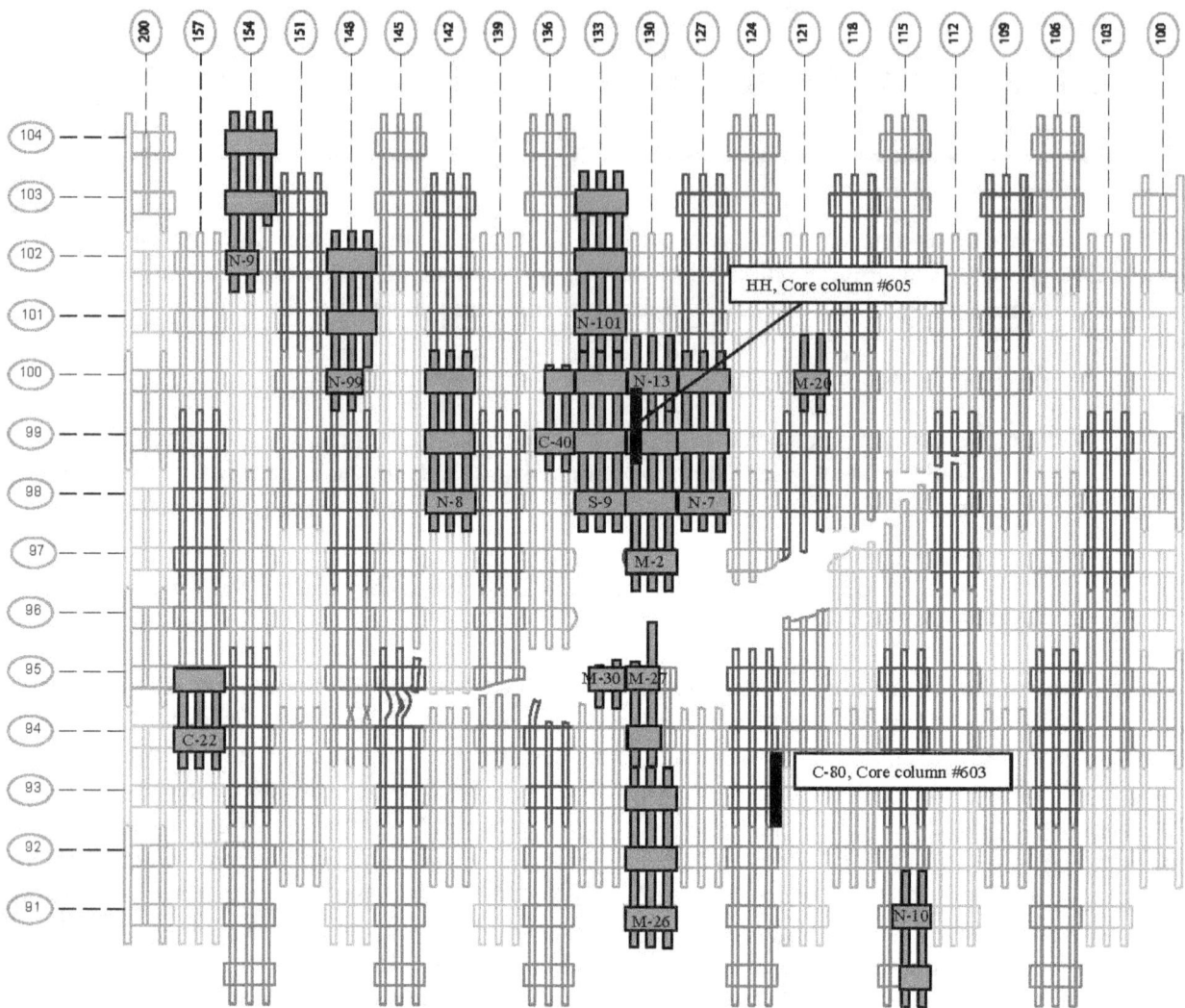

Figure 4–1. Interpreted column damage, from photographic evidence, to WTC 1, with overlay of samples in NIST's possession. Samples shown represent recovered portions. Core columns 603 and 605 are in the second row from the north face of WTC 1.

Structural Steel Elements of Special Importance

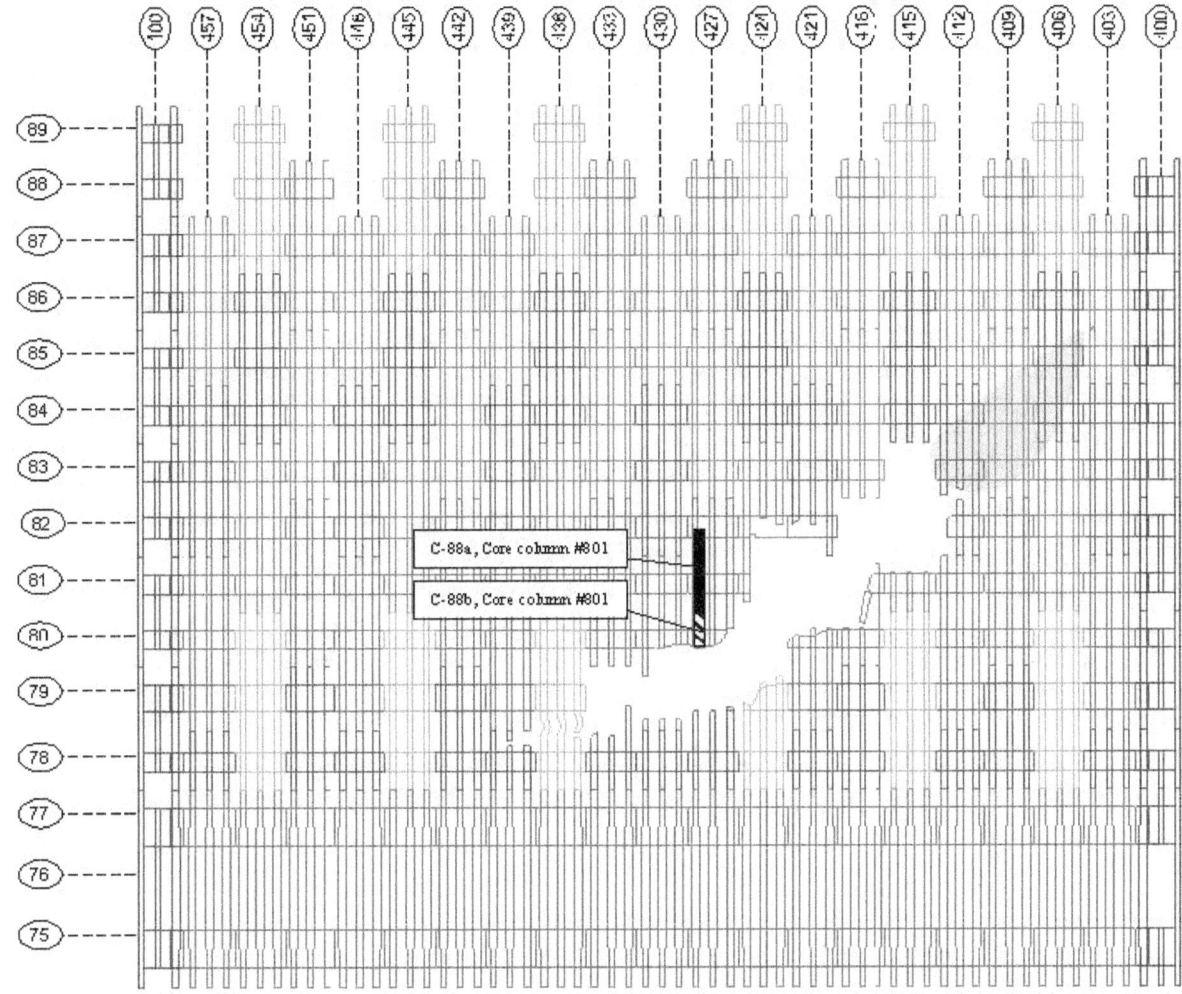

Damage in shaded area cannot be accurately determined.

Note: Core column 801 is in the closest row to the south face of WTC 2.

Figure 4–2. Interpreted column damage, from photographic evidence, to WTC 2, with overlay of samples in NIST's possession.

4.2 SAMPLES REPRESENTING THE VARIOUS TYPES OF STEEL SPECIFIED IN THE DESIGN DRAWINGS

The other grouping of samples that was deemed important was that which belonged to one of the different grades of steel specified in the buildings' construction. The following minimum yield strengths, in ksi (1 ksi equals 1,000 pounds per square inch), were specified for each structural element:

- Columns of the exterior panels: 36, 42, 45, 46, 50, 55, 60, 65, 70, 75, 80, 85, 90, and 100

- Core columns: 36, 42, 46, and 50

- Floor truss material: 36 and 50

From the recovered steel, sufficient representative samples from each important class of steels are available for a full examination (i.e., chemical, metallurgical, and mechanical property analyses). From Table 4–1, it can be seen that 10 of the 14 types of steel specified for the columns are represented, and 10 of the 12 grades of spandrel material have been identified. Additionally, sample ASCE-3 (as-built location in the building not identified) has a flange stamping of 45 for the minimum yield requirement, which would increase the total number of perimeter column material types to 11. One important note is that from the observed stampings of the recovered elements and other documents (see NIST NCSTAR 1-2B), it appears that 100 ksi steel was substituted for the 85 ksi and 90 ksi grades in the construction of the exterior panels (Table 3–6). Considering both column and spandrel material, samples of all grades specified for the perimeter panels are available for metallurgical and mechanical property evaluation. There are a total of 106 individual perimeter columns (97 columns with known as-built locations and 9 columns from unidentified panels sections where the column type and minimum strength values could be deciphered from the stampings located on the base of the columns) and 87 pieces of spandrel material. Tables 4–2 and 4–3 list the specified minimum yield strength/gauge combinations recovered for the columns and spandrels, respectively. While only two of the four grades of steels were obtained (36 ksi and 42 ksi) for the core columns (Table 3–3), 99 percent of the total number of core columns were fabricated from these two grades. For the floor truss material, the samples could not be identified as to their precise, as-built locations within the buildings. However, initial chemical and mechanical property analyses have shown that both minimum yield strength materials specified have been recovered.

Table 4–1. Listing of recovered exterior column panels with specified minimum yield strengths and thickness for columns[a] and spandrels.

NIST NAME	Bldg	Center Col. #	Splice at floor Lower	Splice at floor Upper	Panel Type	COLUMN 1 Col. type	COLUMN 1 Fy (ksi)	COLUMN 2 Col. type	COLUMN 2 Fy (ksi)	COLUMN 3 Col. type	COLUMN 3 Fy (ksi)	LOWER SPANDREL Gage (in)	LOWER SPANDREL Fy (ksi)	MIDDLE SPANDREL Gage (in)	MIDDLE SPANDREL Fy (ksi)	UPPER SPANDREL Gage (in)	UPPER SPANDREL Fy (ksi)
B-1024	WTC 2	154	21	24	300	152	50	150	50	149	50	1.25	36	1.25	36	1.25	36
ASCE-2	WTC 2	330	40	43	400	356	50	356	50	356	50	0.938	36	n/a	n/a	0.938	45
CC	WTC 1	124	70	73	300	133	50	133	50	133	50	0.5625	36	0.5625	36	0.5625	36
M-26	WTC 1	130	90	93	300	125	50	125	55	125	50	0.375	36	0.375	36	0.375	36
M-27	WTC 1	130	93	96	300	123	55	123	55	124	50	0.375	36	0.375	36	0.375	36
M-2	WTC 1	130	96	99	300	122	55	122	55	122	55	0.375	36	0.375	42	0.375	36
M-30	WTC 1	133	94	97	300	123	55	123	55	123	55	0.375	36	0.375	36	0.375	42
C-18	WTC 2	230	93	96	300	120	55	120	55	120	55	0.375	45	0.375	42	0.375	42
N-9	WTC 1	154	101	104	300	120	55	120	55	120	55	0.375	42	0.375	36	0.375	36
M-20	WTC 1	121	99	102	300	120	55	120	55	120	55	0.375	42	0.375	42	0.375	36
N-13	WTC 1	130	99	102	300	120	55	121	55	121	55	0.375	42	0.375	42	0.375	36
N-101	WTC 1	133	100	103	300	120	55	120	55	120	55	0.375	42	0.375	36	0.375	36
S-9	WTC 1	133	97	100	300	122	55	122	55	122	55	0.375	36	0.375	42	0.375	36
N-10	WTC 1	115	89	92	300	125	55	125	55	125	55	0.375	36	0.375	42	0.375	42
C-40	WTC 1	136	98	101	300	121	55	121	60	121	60	0.375	42	0.375	36	0.375	42
C-89	WTC 2	215	12	15	300	143	55	145	50	147	50	1.375	36	1.375	36	1.375	36
N-7	WTC 1	127	97	100	300	121	60	121	55	121	55	0.375	42	0.375	42	0.375	42
C-92	WTC 2	130	93	96	300	123	60	123	60	124	60	0.375	42	0.375	42	0.375	42
C-93	WTC 1	339	99	102	300	121	60	121	60	121	60	0.375	42	0.375	42	0.375	42
K-1	WTC 1	209	97	100	300	120	60	120	60	120	60	0.375	42	0.375	42	0.375	42
K-2	WTC 1	236	92	95	300	120	60	120	60	120	60	0.375	42	0.375	42	0.375	42
N-8	WTC 1	142	97	100	300	121	60	121	60	121	60	0.375	42	0.375	42	0.375	42
C-48	WTC 2	442	91	94	300	120	65	120	65	120	65	0.375	45	0.375	45	0.375	42
N-99	WTC 1	148	99	102	300	120	65	120	65	120	65	0.375	45	0.375	42	0.375	42
S-14	WTC 2	218	91	94	300	120	70	120	65	120	65	0.375	46	0.375	45	0.375	45
M-28	WTC 2	345	98	101	300	120	70	120	70	120	70	0.375	45	0.375	45	0.375	45
C-55	WTC 1	209	94	97	300	120	70	120	70	120	70	0.375	46	0.375	45	0.375	45
S-10	WTC 2	224	92	95	300	120	70	120	70	120	70	0.375	50	0.375	46	0.375	45
S-1	WTC 2	433	79	82	300	123	70	123	70	123	70	0.4375	50	0.4375	46	0.4375	45
N-1	WTC 1	218	82	85	300	123	75	123	75	123	70	0.4375	50	0.375	50	0.375	50
C-46	WTC 2	157	68	71	300	129	65	128	70	126	80	0.625	65	0.625	65	0.5625	65
N-12	WTC 1	206	92	95	300	120	75	120	75	120	75	0.375	50	0.375	50	0.375	46
C-22	WTC 1	157	93	96	300	120	80	120	75	120	80	0.375	65	0.375	60	0.375	60
C-25	WTC 1	206	89	92	300	120	80	120	80	120	80	0.375	55	0.375	55	0.375	55
B-1044	WTC 2	409	40	43	400	335	85	335	80	335	85	0.9375	60	n/a	n/a	0.9375	50
M-10a	WTC 2	209	82	85	300	120	85	120	85	120	85	0.4375	60	0.375	60	0.375	60
M-10b	WTC 2	206	83	86	300	120	85	120	85	120	85	0.375	60	0.375	60	0.375	55
C-10	WTC 1	451	85	88	300	120	85	120	85	120	90	0.375	60	0.375	60	0.375	60
B-1043	WTC 2	406	40	43	400	334	90	334	90	335	85	0.9375	65	n/a	n/a	0.9375	50
C-24	WTC 2	203	74	77	400	325	100	325	100	325	100	0.5624	70	n/a	n/a	0.5625	80
C-13, 13a	WTC 2	200	90	92	210	120	100	520	100	120	100	n/a	n/a	0.375	70	0.375	70
C-14	WTC 2	300	85	87	210	120	100	522	100	122	100	n/a	n/a	0.375	75	0.375	75

a. Columns 1, 2, and 3 are left to right viewed from inside the building.
Note: Strike through indicates section is missing.

Table 4–2. Strength/gauge combinations of perimeter columns recovered by NIST.

Flange F_y (ksi)	Flange Gauge (in.)	Number of Columns Recovered and Identified by NIST
45	1.75	1
50	0.5	2
50	0.5625	2
50	1.0625	2
50	1.8105	1
50	2.0625	1
50	2.125	1
50	2.25	1
50	2.5	1
50	2.625	1
55	0.25	12
55	0.3125	5
55	0.375	6
55	0.4375	3
55	0.5625	3
55	1.375	1
55	1.6875	1
60	0.25	5
60	0.3125	6
60	0.375	1
60	0.5	1
65	0.25	7
65	0.375	1
65	0.8125	1
70	0.25	7
70	0.4375	2
70	0.75	1
75	0.25	3
75	0.4375	2
80	0.25	3
80	0.625	1
80	1.1875	1
85 – 100	0.25	12
85 – 100	0.5625	3
85 – 100	1.125	2
85 – 100	1.1875	3

Table 4–3. Strength/gauge combinations of spandrels recovered by NIST.

Spandrel F_y (ksi)	Spandrel Gauge (in.)	Number of Spandrels Recovered by NIST
36	3/8	16
36	9/16	3
36	1 1/4	3
36	1 3/8	3
42	3/8	24
45	3/8	7
46	3/8	4
50	3/8	5
50	7/16	2
50	15/16	2
55	3/8	2
60	3/8	6
60	15/16	1
65	3/8	1
65	9/16	1
65	5/8	2
65	15/16	1
70	3/8	2
75	3/8	1
80	9/16	1

Chapter 5
SUMMARY

The National Institute of Standards and Technology has 236 samples from the World Trade Center (WTC) buildings, the majority belonging to WTC 1 and WTC 2. These samples represent roughly a half percent of the 200,000 tons of structural steel used in the construction of the two towers. The collection of steel from the WTC towers is sufficient for determining the quality of the steel and for determining mechanical properties as input to models of building performance as sufficient representative samples exist for all 12 grades of exterior panel material, 2 grades of the core column material (that represents 99 percent, by total number, of columns), and both grades for the floor truss material.

Chapter 6
REFERENCES

Astaneh-Asl, A. 2002. *World Trade Center Post-Disaster Reconnaisance and Perishable Structural Engineering Data Collection*, Final Report to National Science Foundation, UCB/CEE-Steel-2002/04. University of California, Department of Civil and Environmental Engineering, Berkeley, CA.

McAllister, T., ed. 2002. *World Trade Center Building Performance Study: Data Collection, Preliminary Observations, and Recommendations.* FEMA 403. Federal Emergency Management Agency. Washington, DC, May.

APPENDIX A
DATA ON RECOVERED WTC STEEL

A.1 DATABASE OF RECOVERED STEEL

Table A–1. List of all WTC steel elements recovered for NIST investigation.

In FEMA report?	NIST Name	Type	Brief Description	Markings	Bldg	Column	Floors	Location
Y	AA	C	2 full columns, thick walled					PL
Y (NSF)	ASCE-2	C	1 full column	B330: 40-43	WTC 2	330	40-43	PL
Y (NSF)	ASCE-3	C	1 column, bottom 1/3rd of left column					PL
	B-1011	RB	Heavy rectangular column	508A: 51-54	WTC 1	508	51-	JFK/PL
	B-1022	W	Thick wide flange with severe bend					205
	B-1024	C	3 full columns	B154: 21-24	WTC 2	154	21-24	JFK
	B-1043	C	Mechanical floor, 3 full columns	B406: 40-43	WTC 2	406	40-43	JFK/PL
	B-1044	C	Mechanical floor, 3 full columns	B409: 40-43	WTC 2	409	40-43	JFK/PL
	B-1044-1	O	Piece of crushed metal decking assoc with B-1044					202
	B-1075	W	Wide flange					205
	B-2150	O	Pieces of aluminum sheathing					202
	B-5004	BT	Bowtie section					JFK/PL
	B-5007	BT	Bowtie section					JFK/PL
	B-6152-1	RB	Heavy rectangular column	803A: 15-18	WTC 1	803	15-19	PL
	B-6152-2	RB	Heavy rectangular column	504A: 33-36	WTC 1	504	33-36	PL
Y	BB	C	Single, thick column					205
Y	C-10	C	Full panel	451: 85-88	WTC 1	451	85-88	PL
Y	C-11	C	2 columns, upper 2/3rds					205
Y	C-13 or S-11	CC	Single rectangular column with large spandrels	B200: 90-92	WTC 2	200	90-92	PL
Y	C-13a or S-19	C	Partial of single column	B200: 90-92	WTC 2	200	90-92	PL
Y	C-14 or S-18	C	1 column, lower 1/3rd	B300: 85-87	WTC 2	300	85-87	B
Y	C-15	C	Partial of single column					205
Y	C-16	C	1 column, upper 1/3rd					205
Y	C-16a	C	Fell off during moving of C-16					205
Y	C-18	C	3 columns, bottom 2/3rds	B230: 93-96	WTC 2	230	93-96	PL
	C-18 Associated	B	One washer and nut					Lab
Y	C-22	C	3 columns, lower 1/2, mangled	A157: 93-96	WTC 1	157	93-96	PL
Y	C-24	C	3 columns, upper 1/2, columns change dimensions	B203: 74-77	WTC 2	203	74-77	205
Y	C-25	C	1 column, lower 1/2	A206: 89-92	WTC 1	206	89-92	PL
Y	C-26	W	Three connected wide flanges	604 & 605 (107) 64 Fy 50				PL
Y	C-28	C	1 column of unknown location					205
Y	C-28B	CC	Corner column, in 2 pieces					205
y	C-29	W	Wide flange					205
Y	C-30 or S-12	W	Wide flange	1008B x04 - 10x	WTC 2	1008	104 - 106	PL
Y	C-31 or S-7	W	Wide flange	676 35				205
Y	C-32	C	1 column, upper 1/3rd					236
Y	C-35	W	Wide flange					205
Y	C-40	C	2 columns, lower 2/3rds	A136: 98-101	WTC 1	136	98-101	PL

Appendix A

Table A–1. List of all WTC steel elements recovered for NIST investigation (continued).

In FEMA report?	NIST Name	Type	Brief Description	Markings	Bldg	Column	Floors	Location
Y	C-41	C	1 column, lower 2/3rds					205
Y	C-42	W	Wide flange					205
Y	C-43	C	1 column, lower 1/2					205
Y	C-44	W	Wide flange, FEMA reported possible core columns	59 S 563				PL
Y	C-45	W	Wide flange, FEMA reported possible core columns	16 S2 563				PL
Y	C-46	C	Nearly 3 full columns	B157: 68-71	WTC 2	157	68-71	PL
Y	C-47	C	3 columns, upper 1/2					236
Y	C-48 or S-5	C	Nearly 2 full columns	B442: 91-94	WTC 2	442	91 - 94	205
Y	C-49 or S-6	C	portion of 1 column					236
Y	C-51	C	2 columns, upper 1/2					205
Y	C-52	C	1 column, upper 2/3rds					205
Y	C-53	J	Floor truss					PL
Y	C-53B	J	Floor truss					PL
Y	C-54	C	1 column, small piece with extended outer web					205
Y	C-55	C	1 column, lower 1/3rd	209: 94-97	WTC 1	209	94-97	PL
Y	C-60	W	Wide flange, S-shaped	193 S1 57				PL
Y	C-61	W	Wide flange	150 S 69				PL
Y	C-62	W	Wide flange	224 (S) <48>				PL
Y	C-64	C	1 column with a lot missing					205
Y	C-65 or S-8	W	Wide flange	904A (86-89) <52>				PL
Y	C-66 or S-17	W	Wide flange					205
Y	C-67	C	1 column, rest unknown					205
Y	C-68	C	1 column, upper 1/2					205
Y	C-69	W	Wide flange					205
Y	C-70	W	Wide flange					205
Y	C-71	W	Wide flange	904A 77-80	WTC 1	904	77 - 80	PL
Y	C-72b	W	Wide flange					205
Y	C-73	C	1 column, upper 1/2					205
Y	C-74	W	Wide flange					205
Y	C-75	C	Portion of 1 column and spandrel, rest unknown					236
Y	C-76	W	Wide flange					205
Y	C-77	C	2 columns from different panels attached at spandrel, 1/3rd of each					205
Y	C-78	W	Wide flange					205
Y	C-79	RB	Rectangular column, FEMA reported possible core column	101A 81 - 85 - 87 -92 52	WTC 1			PL
Y	C-80	W	Wide flange, FEMA reported possible core columns	603A 92-95 <51>	WTC 1	603	92-95	PL
Y	C-81	W	Wide flange					205
Y	C-82	W	Wide flange					205
Y (NSF)	C-83	RB	Heavy rectangular column, FEMA reported as possible core column	No ID, similar to other core column				PL
Y (NSF)	C-84	C	1 full column					PL
Y (NSF)	C-85	W	Wide flange					205
Y	C-87	W	Thick Wide flange					205
Y	C-88a	RB	Built-up box column, FEMA reported possible core column	801B 80-83	WTC 2	801	80-83	PL
	C-88b		Built-up box column, welded to C-88a	801B 77-80	WTC 2	801	77-80	PL
	C88c	O	Broke off C-88					PL
Y (NSF)	C-89	C	2 full columns	B 215: 12 - 15	WTC 2	215	12 - 15	PL

Table A–1. List of all WTC steel elements recovered for NIST investigation (continued).

In FEMA report?	NIST Name	Type	Brief Description	Markings	Bldg	Column	Floors	Location
Y (NSF)	C-90	RB	Heavy rectangular column, FEMA reported as possible core column	701B 12 - 15	WTC 2	701	12 - 15	PL
Y	C-91	Ch	Channel					236
Y	C-92	C	Partial of single column	B13x: 93-96	WTC 2	130	93 - 96	PL
Y	C-93	C	Partial of single column	339: 99 - 102	WTC 1	339	99 - 102	PL
	C-94	O	May be some type of brace, rectangular box construction					PL
	C-95	Ch	Channel					236
	C-96	Ch	Channel					236
	C-97	Ch	Channel					236
	C-98	Ch	Channel					236
	C-99	Ch	Channel					236
	C-100	J	Possible angle from a floor truss					PL
	C-101	RB	thinner	78A 10 27 50				PL
	C-102	C	Partial of single column					205
	C-103	O	Square-tube construction					PL
	C-104	J	Possible angle from a floor truss					PL
	C-105	Ch	Channel					236
	C-106	J	Small piece of floor truss					202
	C-107	Ch	Channel					236
	C-108	B	Three sheared bolts					Lab
	C-109	B	Single bolt sheared					Lab
	C-110	B	Bolt and nut					Lab
	C-111	B	Bolt and washer					Lab
	C-112	B	Single bolt sheared					Lab
	C-113	B	Two sheared bolts with washers					Lab
	C-114	B	Sheared bolt with nut					Lab
	C-115	J	Pig-tailed piece from floor truss					Lab
	C-116	H	Damper					Lab
	C-117	C	3 columns, lower 1/3	101-104				PL
	C-118	Ch	Channel					236
	C-119A	O	Square-tube construction					PL
	C-119B	O	Square-tube construction					PL
	C-120	O	Square-tube construction					PL
	C-121	O	Square-tube construction					PL
	C-122	J	Piece of floor truss					PL
	C-123	W	Small Wide flange					205
	C-124	Ch	Channel					236
	C-125	Ch	Channel					236
	C-126	W	Wide flange					205
	C-128	Ch	Channel					B
	C-129	Ch	Channel					236
	C-130	W	Wide Flange					205
	C-131	J	Small portion of floor truss with cement					202
	C-132	J	Piece of floor truss					PL
	C-133	C	1 column, bottom 1/3rd of unknown location					205
	C-134	Ch	Channel					236
	C-135	O	May be some type of brace, rectangular box construction					PL
	C-137a	J	Piece of floor truss					PL
	C-137b	J	Piece of floor truss					PL
	C-137c	J	Piece of floor truss					PL
	C-137d	J	Piece of floor truss					PL
	C-137f	J	Piece of floor truss					PL

Appendix A

Table A–1. List of all WTC steel elements recovered for NIST investigation (continued).

In FEMA report?	NIST Name	Type	Brief Description	Markings	Bldg	Column	Floors	Location
	C-138	W	Small wide flange					205
	C-139	Ch	Channel					236
	C-140	J	Piece of angle					PL
	C-141	Ch	Channel					236
	C-142	W	Wide flange					205
	C-143	Ch	Channel					236
	C-144	Ch	Channel					236
	C-145	Ch	Channel					236
	C-146a	O	Mangled ball of steel and concrete					202
	C-146b	J	Piece of floor truss					PL
	C-147	Ch	Channel					236
	C-148	Ch	Channel					236
	C-149	J	Piece of floor truss					PL
	C-150	W	Wide flange					205
	C-151	J	Piece of floor truss					PL
	C-152	Ch	Channel					236
	C-153	Ch	Channel					236
	C-154	RB	Thin rectangular beam with supports	825: 107-108 52				PL
	C-155	W	Wide flange	904A 83-86	WTC 1	904	83-86	PL
	C-156	O	Square-tube construction					PL
Y	CC	C	2 full columns	124: 73-70	WTC 1	124	70-73	PL
Y	DD	C	1 Column, spans 1 floor and has end plates on both ends					205
Y	FF	C	Single, thick column					205
	GZ-1	Cn5	Received from D. Sharp, coupon from Bldg #5					Lab
	GZ-2	Cn5	Received from D. Sharp, coupon from Bldg #5					Lab
	GZ-3	Cn5	Received from D. Sharp, coupon from Bldg #5					Lab
	GZ-4	Cn5	Received from D. Sharp, coupon from Bldg #5					Lab
	GZ-5	Cn5	Received from D. Sharp, coupon from Bldg #5					Lab
	GZ-6	Cn5	Received from D. Sharp, coupon from Bldg #5					Lab
	GZ-7	Cn5	Received from D. Sharp, coupon from Bldg #5					Lab
Y	HH or S-2	W	Wide flange, FEMA reported possible core column	605A 98-101	WTC 1	605	98-101	PL
Y	K-1 or K-13	C	3 columns, lower 1/3rd	209: 97-100	WTC 1	209	97-100	202
Y	K-2 or K-40	C	1 column, lower 2/3rds	236: 92-95	WTC 1	236	92-95	PL
Y	K-10	Cn	Flange coupon received from Gross, July 29, 2002					Lab
Y	K-11	Cn	Flange coupon received from Gross, July 29, 2002					Lab
Y	K-12	Cn	Flange coupon received from Gross, July 29, 2002					Lab
Y	K-13	Cn	Flange coupon received from Gross, July 29, 2002					Lab

Table A–1. List of all WTC steel elements recovered for NIST investigation (continued).

In FEMA report?	NIST Name	Type	Brief Description	Markings	Bldg	Column	Floors	Location
Y	K-14	Cn	Flange coupon received from Gross, July 29, 2002					Lab
Y	K-15	Cn	Flange coupon received from Gross, July 29, 2002					Lab
Y	K-16	C	1 full column, thick, looks very corroded					PL
	K-16a	C	Fell off of K-16 while moving					PL
Y	K-18	Cn	Flange coupon received from Gross, July 29, 2002					Lab
Y	K-19a	Cn	Flange coupon received from Gross, July 29, 2002					Lab
Y	K-19b	Cn	Flange coupon received from Gross, July 29, 2002					Lab
Y	K-50a	O	Rectangular slab of steel with bolts, received from D. Sharp, SEAoNY					Lab
Y	K-50b	O	Rectangular slab of steel with bolts, received from D. Sharp, SEAoNY					Lab
Y	K-50c	O	Rectangular slab of steel with bolts, received from D. Sharp, SEAoNY					Lab
Y	M-2	C	Full panel	-9 <63>	WTC 1	130	96-99	PL
Y	M-4 or M-5	C	3 columns, upper 2/3rds					205
Y	M-10a	C	3 columns, unknown location	B209: 82-85	WTC 2	206	82-85	PL
Y	M-10b	C	3 columns, lower 1/2	B206: 83-86	WTC 2	206	83-86	PL
Y	M-11	W	Wide flange					205
Y	M-17	W	Wide flange or I-beam, 1' flange, 2' web, 50-60' long	163 (9) 62				205
	M-17a	O	Fell off of M-17 while moving					202
	M-18	RB	Large box beam, 19" x 21" x 17.5' long					205
	M-19	C	2 columns, upper 1/3rd					205
	M-20	C	2 columns, lower 1/3rd	A121: 99-102	WTC 1	121	99-102	PL
	M-22	RB	Large box beam, 19" x 26.5" x 9.5' long					205
	M-23	W	Possibly part of Wide flange or I-beam	F 2010				PL
	M-24	Ch	Channel					236
	M-25	J	Small piece of floor truss					202
	M-26	C	3 full columns	A130: 90-93	WTC 1	130	90-93	PL
	M-26 associated	B	8 bolts and a nut					Lab
	M-27	C	2 columns, lower 3/4ths	A130: 93-96	WTC 1	130	93-96	202
	M-28	C	3 columns, lower 1/4th	B345: 9x - 1xx	WTC 2	345	98 - 101	PL
	M-29	O	5 ft piece of strapping					202
	M-30	C	2 columns, lower 1/3rd	_33: 94-97	WTC 1	133	94-97	202
	M-30 associated	O	Pieces of glass, plexiglass, other rubble					Lab
	M-31	J	Pieces of floor truss					Lab
	M-32	J	Pieces of floor truss					Lab
	M-33	W	Wide flange					205
	M-34	Ch	Channel					B
	M-35	CC	Corner column					205
	M-36	J	Thick angle					PL
	M-37	W	Wide flange	130 (8?–92) <50>				205
	M-38	W	Wide flange	Fy 42				PL

Appendix A

Table A–1. List of all WTC steel elements recovered for NIST investigation (continued).

In FEMA report?	NIST Name	Type	Brief Description	Markings	Bldg	Column	Floors	Location
Y	N-1	C	2 full columns	2_8: 82-85	WTC 1	218	82-85	PL
Y	N-3	C	1 column, upper 1/2					236
Y	N-4	C	1 column, middle 1/3rd					236
Y	N-5	O	Part of spandrel plate with bolts					PL
Y	N-6	C	1 column, length of spandrel, crushed					236
Y (as M-3)	N-7 or M-3	C	3 full columns	127: 97-100	WTC 1	127	97-100	PL
Y (as M-7)	N-8 or M-7	C	Full panel	A142: 97-100	WTC 1	142	97-100	PL
Y (as M-8)	N-9 or M-8	C	Almost full panel, missing lower 1/3rd of 1 column	A154: 101-104	WTC 1	154	101-104	PL
15)	N-10 or M-15	C	2 columns, lower 2/3rds	A115: 89-92	WTC 1	115	89-92	PL
Y (as M-9)	N-11 or M-9	C	3 columns, upper 2/3rds					205
13)	N-12 or M-13	C	2 full columns	_06: 92-95	WTC 1	206	92-95	PL
14)	N-13 or M-14	C	3 columns, lower 1/3rd	A130: 99-102	WTC 1	130	99-102	B
Y (as M-16)	N-99 or M-16	C	Almost full panel, missing lower 1/3rd of 1 column	A148: 99-102	WTC 1	148	99-102	PL
	N-101 or M-21	C	3 full columns	A133: 100-103	WTC 1	133	100-103	PL
Y (as C-19)	N-N or C-19	C	1 column, lower 1/2					205
Y (as EE)	S-1 or EE	C	2 columns, lower 1/3rd	A433: 79-82	WTC 1	433	79-82	PL
Y (as C-50)	S-3 or C-50	C	1 column, unknown 1/2					205
Y (as C-63)	S-9 or C-63	C	Full panel	A133: 97-100	WTC 1	133	97-100	PL
Y (as C-17)	S-10 or C-17	C	2 columns, lower 1/2	224: 92-95	WTC 1	224	92-95	PL
Y (as C-20)	S-14 or C-20	C	Full panel	B218: 91-94	WTC 2	218	91-94	PL
	SM-2	W	I-beam					205
Y (as N-2)	T-1 or N-2	J	Floor truss material					202
	U-6	C	3 columns, upper 1/4					236
	U-15	C	Partial of single column					205
	U-25	O	Unknown Wide flange with concrete	<North> 84-155 A8 Div 2				205
Y	W-14A or A	W	Heavy Wide flange					205
Y	W-14B	W	Heavy Wide flange					PL

NSF: Pieces contributed by A. Asteneh salvaged under NSF contract

Key: 202, Bldg. 202, high bay; 205, Bldg. 205, parking lot; 236, Bldg. 236, parking lot; B, bolt; BT, bowtie section of exterior wall; C, flat wall, exterior column panel section; CC, corner panel section of exterior wall; Ch, channel; Cn, coupon of exterior column; Cn5, coupon from WTC 5; H, hanger; J, floor truss; NSF, pieces contributed by A. Asteneh salvaged under NSF contract; O, other; RB, rectangular, built-up box column; W, wide flange section; Lab, Bldg. 223, Rm B253; JFK, Hanger 17, JFK Airport; JFK/PL, Main piece at JFK, portion at NIST. PL, Bldg. 202, parking lot;

Table A–2. List of identified exterior panel sections.

In FEMA report?	NIST Name	Type	Brief Description	Markings	Bldg	Column	Floors
Y (NSF)	ASCE-2	C	1 full column	B330: 40-43	WTC 2	330	40-43
	B-1024	C	3 full columns	B154: 21-24	WTC 2	154	21-24
	B-1043	C	Mechanical floor, 3 full columns	B406: 40-43	WTC 2	406	40-43
	B-1044	C	Mechanical floor, 3 full columns	B409: 40-43	WTC 2	409	40-43
Y	C-10	C	Full panel	451: 85-88	WTC 1	451	85-88
Y	C-13 or S-11	CC	Single rectangular column with large spandrels	B200: 90-92	WTC 2	200	90-92
Y	C-13a or S-19	C	Partial of single column	B200: 90-92	WTC 2	200	90-92
Y	C-14 or S-18	C	1 column, lower 1/3rd	B300: 85-87	WTC 2	300	85-87
Y	C-18	C	3 columns, bottom 2/3rds	B230: 93-96	WTC 2	230	93-96
Y	C-22	C	3 columns, lower 1/2, mangled	A157: 93-96	WTC 1	157	93-96
Y	C-24	C	3 columns, upper 1/2, columns change dimensions	B203: 74-77	WTC 2	203	74-77
Y	C-25	C	1 column, lower 1/2	A206: 89-92	WTC 1	206	89-92
Y	C-40	C	2 columns, lower 2/3rds	A136: 98-101	WTC 1	136	98-101
Y	C-46	C	Nearly 3 full columns	B157: 68-71	WTC 2	157	68-71
Y	C-48 or S-5	C	Nearly 2 full columns	B442: 91-94	WTC 2	442	91 - 94
Y	C-55	C	1 column, lower 1/3rd	209: 94-97	WTC 1	209	94-97
Y (NSF)	C-89	C	2 full columns	B215: 12 - 15	WTC 2	215	12 - 15
Y	C-92	C	Partial of single column	B13x: 93-96	WTC 2	130	93 - 96
Y	C-93	C	Partial of single column	339: 99 - 102	WTC 1	339	99 - 102
Y	CC	C	2 full columns	124: 73-70	WTC 1	124	70-73
Does not match	K-1 or K-13	C	3 columns, lower 1/3rd	209: 97-100	WTC 1	209	97-100
Y	K-2 or K-40	C	1 column, lower 2/3rds	236: 92-95	WTC 1	236	92-95
Y	M-2	C	Full panel	-9 <63>	WTC 1	130	96-99
	M-10a	C	3 columns, 1/3rd, not labeled but attached to M-10b	B209: 82-85	WTC 2	209	82-85
Y	M-10b	C	3 columns, lower 1/2	B206: 83-86	WTC 2	206	83-86
	M-20	C	2 columns, lower 1/3rd	A121: 99-102	WTC 1	121	99-102
	M-26	C	3 full columns	A130: 90-93	WTC 1	130	90-93
	M-27	C	2 columns, lower 3/4ths	A130: 93-96	WTC 1	130	93-96
	M-28	C	3 columns, lower 1/4th	B345: 9x - 1xx	WTC 2	345	98 - 101
	M-30	C	2 columns, lower 1/3rd	_33: 94-97	WTC 1	133	94-97
Y	N-1	C	2 full columns	2_8: 82-85	WTC 1	218	82-85
Y (as M-3)	N-7 or M-3	C	3 full columns	127: 97-100	WTC 1	127	97-100
Y (as M-7)	N-8 or M-7	C	Full panel	A142: 97-100	WTC 1	142	97-100
Y (as M-8)	N-9 or M-8	C	Almost full panel, missing lower 1/3rd of 1 column	A154: 101-104	WTC 1	154	101-104
Y (as M-15)	N-10 or M-15	C	2 columns, lower 2/3rds	A115: 89-92	WTC 1	115	89-92
Y (as M-13)	N-12 or M-13	C	2 full columns	_06: 92-95	WTC 1	206	92-95
Y (as M-14)	N-13 or M-14	C	3 columns, lower 1/3rd	A130: 99-102	WTC 1	130	99-102
	N-99 or M-16	C	Almost full panel, missing lower 1/3rd of 1 column	A148: 99-102	WTC 1	148	99-102
	N-101 or M-21	C	3 full columns	A133: 100-103	WTC 1	133	100-103
	S-1 or EE	C	2 columns, lower 1/3rd	A433: 79-82	WTC 1	433	79-82
Y	S-9 or C-63	C	Full panel	A133: 97-100	WTC 1	133	97-100
Y	S-10 or C-17	C	2 columns, lower 1/2	224: 92-95	WTC 1	224	92-95
Y	S-14 or C-20	C	Full panel	B218: 91-94	WTC 2	218	91-94

Table A–3. List of partially identified exterior panel sections.

In FEMA report?	NIST Name	Type	Brief Description	Markings	Bldg	Column	Floors
	C-117	C	3 columns, lower 1/3	101-104	NA		101-104

Appendix A

Table A–4. List of unidentified exterior panel sections.

In FEMA report?	NIST Name	Type	Brief Description	Location
Y	C-28B (formerly U-4)	CC	Corner column, in 2 pieces	205
	M-35	CC	Corner column	205
Y	AA (formerly U-7)	C	2 full columns, thick walled	PL
Y (NSF)	ASCE-3	C	1 column, bottom 1/3rd of left column	PL
Y	BB	C	Single, thick column	205
Y	C-11	C	2 columns, upper 2/3rds	205
Y	C-15 (formerly U-20)	C	Partial of single column	205
Y	C-16	C	1 column, upper 1/3rd	205
Y	C-16a	C	Fell off during moving of C-16	205
Y	C-28 (formerly U-1)	C	1 column of unknown location	205
Y	C-32	C	1 column, upper 1/3rd	236
Y	C-41	C	1 column, lower 2/3rds	205
Y	C-43	C	1 column, lower 1/2	205
	C-47	C	3 columns, upper 1/2	236
Y	C-49 or S-6	C	portion of 1 column	236
Y	C-51	C	2 columns, upper 1/2	205
Y	C-52	C	1 column, upper 2/3rds	205
Y	C-54	C	1 column, small piece with extended outer web	205
Y	C-64	C	1 column with a lot missing	205
Y	C-67	C	1 column, rest unknown	205
Y	C-68	C	1 column, upper 1/2	205
Y	C-73	C	1 column, upper 1/2	205
Y	C-75	C	portion of 1 column and spandrel, rest unknown	236
Y	C-77	C	2 columns from different panels attached at spandrel, 1/3rd of each	205
Y (NSF)	C-84	C	1 full column, stampings on front face	PL
	C-102	C	Partial of single column	205
	C-133	C	1 column, bottom 1/3rd of unknown location	205
Y	DD	C	1 Column, spans 1 floor and has end plates on both ends	205
Y	FF	C	Single, thick column	205
Y	K-16	C	1 full column, thick, looks very corroded	PL
	K-16a (formerly U-23)	C	Fell off of K-16 while moving	PL
Both are in report but listed separately	M-4 or M-5	C	3 columns, upper 2/3rds	205
	M-19	C	2 columns, upper 1/3rd	205
Y	N-3	C	1 column, upper 1/2	236
Y	N-4	C	1 column, middle 1/3rd	236
Y	N-6 (formerly U-2)	C	1 column, length of spandrel, crushed	236
Y (as M-9)	N-11 or M-9	C	3 columns, upper 2/3rds	205
Y (as C-19)	N-N or C-19	C	1 column, lower 1/2	205
Y (as C-50)	S-3 or C-50	C	1 column, unknown 1/2	205
	U-6	C	3 columns, upper 1/4	236
	U-15	C	Partial of single column	205
Y	K-10	Cn	Flange coupon received from Gross, July 29, 2002	Lab
Y	K-11	Cn	Flange coupon received from Gross, July 29, 2002	Lab
Y	K-12	Cn	Flange coupon received from Gross, July 29, 2002	Lab
Y	K-13	Cn	Flange coupon received from Gross, July 29, 2002	Lab
Y	K-14	Cn	Flange coupon received from Gross, July 29, 2002	Lab
Y	K-15	Cn	Flange coupon received from Gross, July 29, 2002	Lab
Y	K-18	Cn	Flange coupon received from Gross, July 29, 2002	Lab
Y	K-19a	Cn	Flange coupon received from Gross, July 29, 2002	Lab
Y	K-19b	Cn	Flange coupon received from Gross, July 29, 2002	Lab
	B-5004	BT	Bowtie section	JFK/PL
	B-5007	BT	Bowtie section	JFK/PL

Table A–5. List of identified core columns.

In FEMA report?	NIST Name	Type	Brief Description	Markings	Bldg	Column	Floors
	B-1011	RB	Heavy rectangular column	508A: 51-54 <55>	WTC 1	508	51-54
	B-6152-1	RB	Heavy rectangular column	803A: 15-18 <52>	WTC 1	803	15-18
	B-6152-2	RB	Heavy rectangular column	504A: 33-36	WTC 1	504	33-36
NSF	C-83	RB	Heavy rectangular column, FEMA reported as possible core column	No ID found, but similar to core column size and shape			
	C-88a	RB	Built-up box column, FEMA reported possible core column	801B 80-83	WTC 2	801	80-83
	C-88b		Similar shape welded to above column	801B 77-80	WTC 2	801	77-80
NSF	C-90	RB	Heavy rectangular column, FEMA reported as possible core column	701B 12 - 15	WTC 2	701	12 - 15
	C-30 or S-12	W	Wide flange	1008B x04 - 10x	WTC 2	1008	104 - 106
	C-65 or S-8	W	Wide flange	904A (86-89) <52>	WTC 1	904	86-89
Y	C-71	W	Wide flange	904A 77-80	WTC 1	904	77 - 80
	C-80	W	Wide flange, FEMA reported possible core columns	603A 92-95 <51>	WTC 1	603	92-95
	C-155 (formerly U-5)	W	Wide flange	904A 83-86	WTC 1	904	83-86
	HH or S-2	W	Wide flange, FEMA reported possible core columns	605A 98-101	WTC 1	605	98-101

Table A–6. List of built-up box beams and wide flange sections with ambiguous stampings.

NIST Name	Type	Brief Description	Markings	Location
Markings but no knowledge of this coding				
C-79	RB	Rectangular column, FEMA reported possible core column	101A 81 - 85 - 87 -92 52	PL
C-101 (formerly U-16)	RB	Similar to corner column, but much thinner	78A 10 27 50	PL
C-154	RB	Thin rectangular beam with supports	825: 107-108 52	PL
C-26	W	Three connected Wide flanges	604 & 605 (107) <64> Fy 50	PL
C-44	W	Wide flange, FEMA reported possible core columns	59 S 563	PL
C-45	W	Wide flange, FEMA reported possible core columns	16 S2 563 Fy 50	PL
C-60	W	Wide flange, S-shaped	193 S1 57	PL
C-61	W	Wide flange	150 S 69	PL
C-62	W	Wide flange	224 (S) <48> Fy 50	PL
M-17	W	Wide flange or I-beam, 1ft flange, 2 ft web, 50-60 ft long	163 (9) 62 Fy 36	205
M-23	W	Possibly part of Wide flange or I-beam	F 2010	PL
M-37	W	Wide flange	130 (8?–92) <50>	205
M-38	W	Wide flange	Fy 42	PL

Table A–7. List of unidentified wide flange sections.

In FEMA report?	NIST Name	Type	Brief Description	Location
	B-1022	W	Thick wide flange with severe bend	205
	B-1075	W	Wide flange	205
Y	C-29 (formerly U-10)	W	Wide flange	205
Y	C-35	W	Wide flange	205
Y	C-69	W	Wide flange	205
Y	C-70 (formerly U-9)	W	Wide flange	205
Y	C-72b	W	Wide flange	205
Y	C-76	W	Wide flange	205
Y	C-78 (formerly U-8)	W	Wide flange	205
Y	C-81	W	Wide flange	205
Y	C-82	W	Wide flange	205
Y (NSF)	C-85	W	Wide flange	205
Y	C-87	W	Thick Wide flange	205
	C-123	W	Small Wide flange	205
	C-126	W	Wide flange	205
	C-130	W	Wide flange	205
	C-138	W	Wide flange	205
	C-142	W	Wide flange	205
	C-150	W	Wide flange	205
Y	M-11	W	Wide flange	205
	M-18	RB	Large box beam	205
	M-22	RB	Large box beam	205
	M-33	W	Wide flange	205
	SM-2	W	Wide flange	205
Y	W-14A or A	W	Heavy Wide flange	205
Y	W-14B	W	Heavy Wide flange	PL

Table A–8. List of recovered floor truss material.

In FEMA report?	NIST Name	Type	Brief Description	Location
Y	C-53	J	Floor truss	PL
Y	C-53B	J	Floor truss	PL
	C-100	J	Possible angle from a foor truss	PL
	C-104	J	Possible angle from a foor truss	PL
	C-106 (formerly U-18)	J	Small piece of floor truss	202
	C-115	J	Pig-tailed piece from flcor truss	Lab
	C-122	J	Piece of floor truss	PL
	C-131	J	Small portion of floor truss with cement	202
	C-132	J	Piece of floor truss	PL
	C-137a	J	Piece of floor truss	PL
	C-137b	J	Piece of floor truss	PL
	C-137c	J	Piece of floor truss	PL
	C-137d	J	Piece of floor truss	PL
	C-137f	J	Piece of floor truss	PL
	C-140	J	Piece of angle	PL
	C-146b	J	Piece of floor truss	PL
	C-149	J	Piece of floor truss	PL
	C-151	J	Piece of floor truss	PL
	M-25	J	Small piece of floor truss	202
	M-31	J	Pieces of floor truss	Lab
	M-32	J	Pieces of floor truss	Lab
	M-36	J	Thick angle from floor truss	PL
Y (as N-2)	T-1 or N-2	J	Floor truss	202

Table A–9. List of recovered channel material.

In FEMA report?	NIST Name	Type	Brief Description	Location
Y	C-91	Ch	Channel	236
	C-95	Ch	Channel	236
	C-96	Ch	Channel	236
	C-97	Ch	Channel	236
	C-98	Ch	Channel	236
	C-99	Ch	Channel	236
	C-105	Ch	Channel	236
	C-107 (formerly U-19)	Ch	Channel	236
	C-118	Ch	Channel	236
	C-124	Ch	Channel	236
	C-125	Ch	Channel	236
	C-128	Ch	Channel	B
	C-129	Ch	Channel	236
	C-134	Ch	Channel	236
	C-139	Ch	Channel	236
	C-141	Ch	Channel	236
	C-143	Ch	Channel	236
	C-144	Ch	Channel	236
	C-145	Ch	Channel	236
	C-147	Ch	Channel	236
	C-148	Ch	Channel	236
	C-152	Ch	Channel	236
	C-153	Ch	Channel	236
	M-24	Ch	Channel	236
	M-34	Ch	Channel	B

Appendix A

Table A–10. List of material from WTC 5.

In FEMA report?	NIST Name	Type	Brief Description	Location
	GZ-1	Cn5	Coupon from Bldg #5	Lab
	GZ-2	Cn5	Coupon from Bldg #5	Lab
	GZ-3	Cn5	Coupon from Bldg #5	Lab
	GZ-4	Cn5	Coupon from Bldg #5	Lab
	GZ-5	Cn5	Coupon from Bldg #5	Lab
	GZ-6	Cn5	Coupon from Bldg #5	Lab
	GZ-7	Cn5	Coupon from Bldg #5	Lab

Table A–11. List of miscellaneous material.

In FEMA report?	NIST Name	Type	Brief Description	Location
	C-18 Associated	B	One washer and nut	Lab
	C-108	B	Three sheared bolts	Lab
	C-109	B	Single bolt sheared	Lab
	C-110	B	Bolt and nut	Lab
	C-111	B	Bolt and washer	Lab
	C-112	B	Single bolt sheared	Lab
	C-113	B	Two sheared bolts with washers	Lab
	C-114	B	Sheared bolt with nut	Lab
	M-26 associated	B	8 bolts and a nut	Lab
	C-116	H	Damper	Lab
	B-1044-1	O	Piece of crushed metal decking assoc with B-1044	202
	B-2150	O	Pieces of aluminum sheathing	202
	C88c (formerly U-22)	O	Broke off C-88	PL
	C-94	O	May be some type of brace, rectangular box construction	PL
	C-103	O	Square-tube construction	PL
	C-119A	O	Square-tube construction	PL
	C-119B	O	Square-tube construction	PL
	C-120	O	Square-tube construction	PL
	C-121	O	Square-tube construction	PL
	C-135	O	May be some type of brace, rectangular box construction	PL
	C-146	O	Mangled ball of steel and concrete	202
	C-156 (formerly U-17)	O	Square-tube construction	PL
Y	K-50a	O	Rectangular slab of steel with bolts, received from D. Sharp, SEAoNY	Lab
Y	K-50b	O	Rectangular slab of steel with bolts, received from D. Sharp, SEAoNY	Lab
Y	K-50c	O	Rectangular slab of steel with bolts, received from D. Sharp, SEAoNY	Lab
	M-17a (formerly U-24)	O	Fell off of M-17 while moving	202
	M-29	O	5 ft piece of strapping	202
	M-30 associated	O	Pieces of glass, plexiglass, other rubble	Lab
Y	N-5	O	Plate with bolts	PL
	U-25	O	Unknown Wide flange with concrete	205

A.2 REPRESENTATIVE PICTURES OF RECOVERED WTC STEEL

Appendix A

Source: NIST.

Figure A–1. Exterior column panel, sample C-46 shown.

Data on Recovered WTC Steel

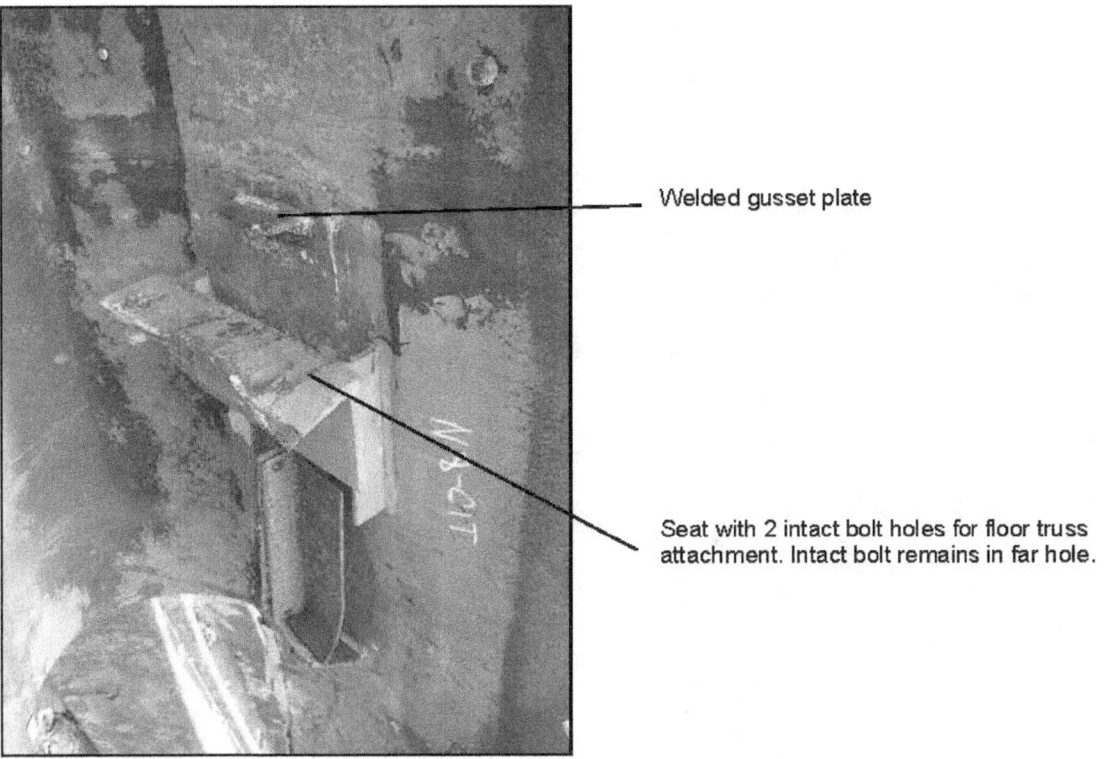

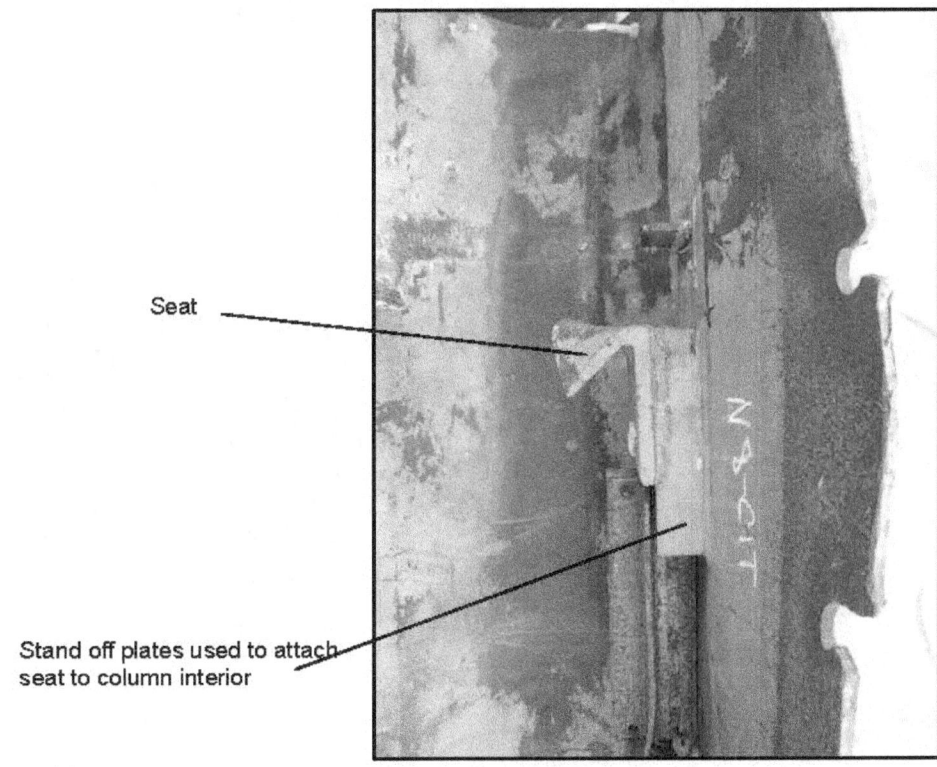

Source: NIST.

Figure A–2. Floor truss seats shown from sample N-8.

Appendix A

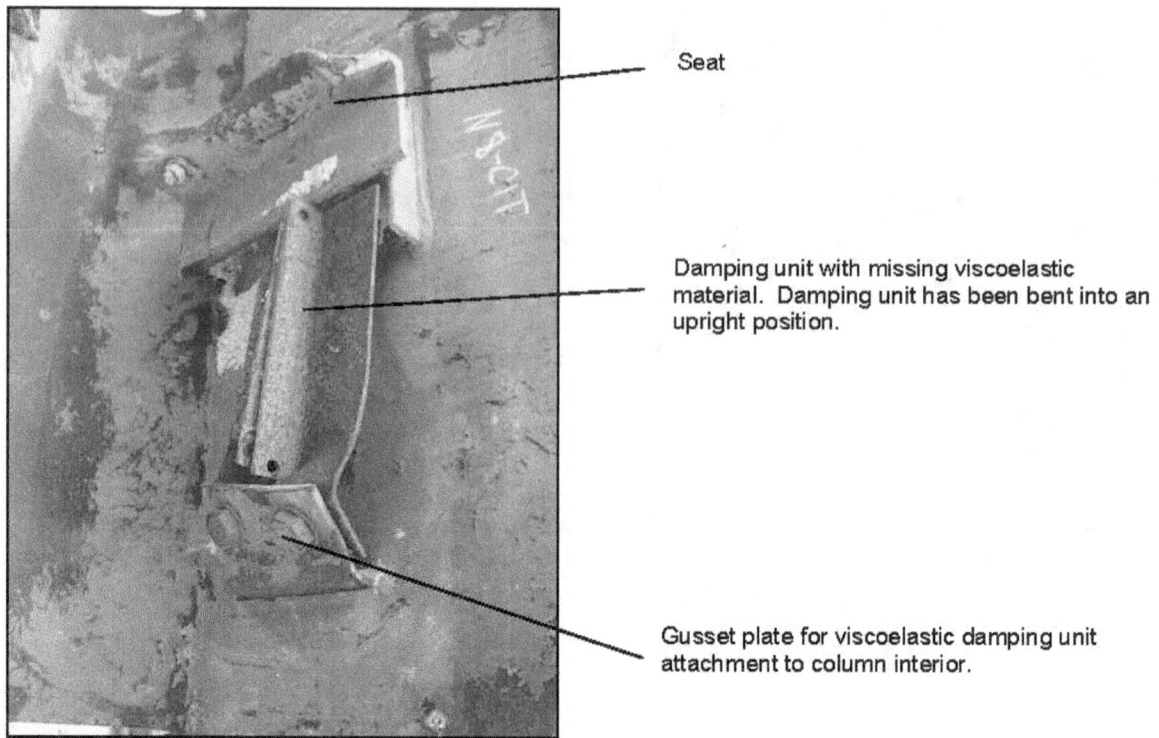

Seat

Damping unit with missing viscoelastic material. Damping unit has been bent into an upright position.

Gusset plate for viscoelastic damping unit attachment to column interior.

Figure A–3. Damping Unit shown from sample N-8.

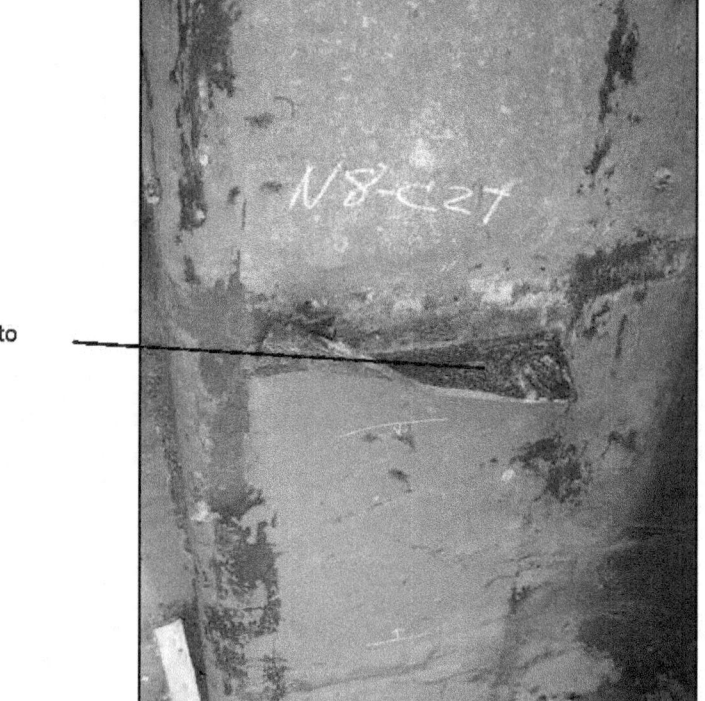

Welded gusset plate used in place of seat on alternate column/spandrel intersections. One method used to attach diagonal bracing strap to exterior wall

Source: NIST.

Figure A–4. Gusset plate shown from sample N-8.

Data on Recovered WTC Steel

Diagonal bracing strap attached directly to exterior column

On Sample C-25

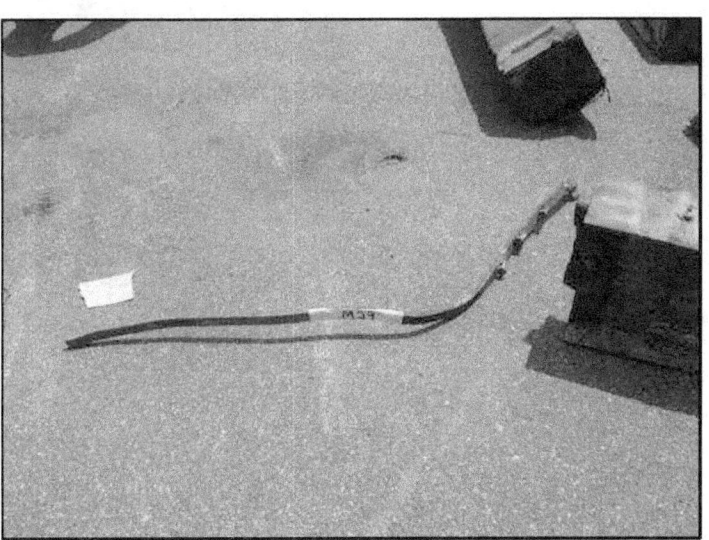

Sample M-29

Source: NIST.

Figure A–5. Diagonal bracing strap shown on sample C-25 (top), and single strap labeled M-29 (bottom).

Appendix A

B-5004 at JFK

B-5004 portion cut and moved to NIST campus

Source: NIST.

Figure A–6. Bowtie section of exterior wall.

Samples C-83 and C-90

Source: NIST.

Figure A–7. Recovered rectangular built up box sections used as core columns.

Appendix A

Sample C-65

Sample C-80

Source: NIST.

Figure A–8. Recovered wide flange sections used as core columns.

Data on Recovered WTC Steel

Source: NIST.

Figure A–9. Other recovered wide flange sections, shown is sample C-42.

Appendix A

Source: NIST.

Figure A–10. Recovered floor truss material; shown are portions of sample C-53.

Source: NIST.

Figure A–11. Recovered inner channel material used to connect floor trusses to core columns; shown is sample C-129.

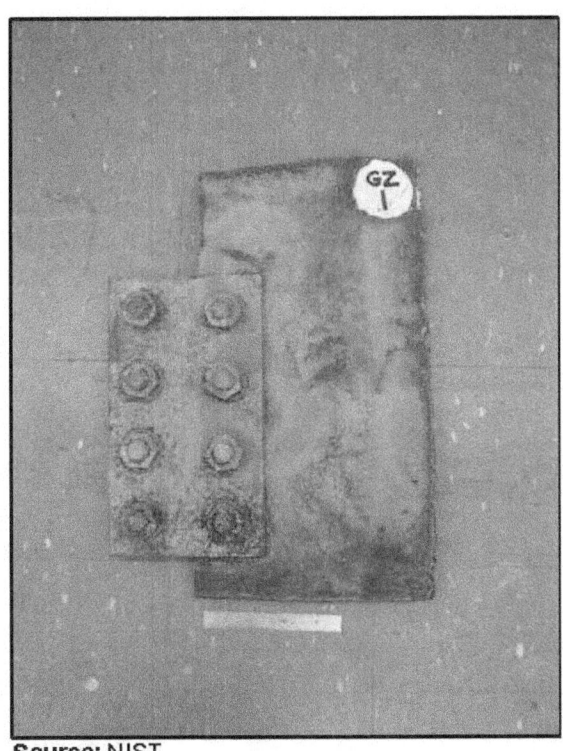 

Source: NIST.

Figure A–12. Coupons removed in the field from WTC 5; shown is sample GZ-1.

Appendix A

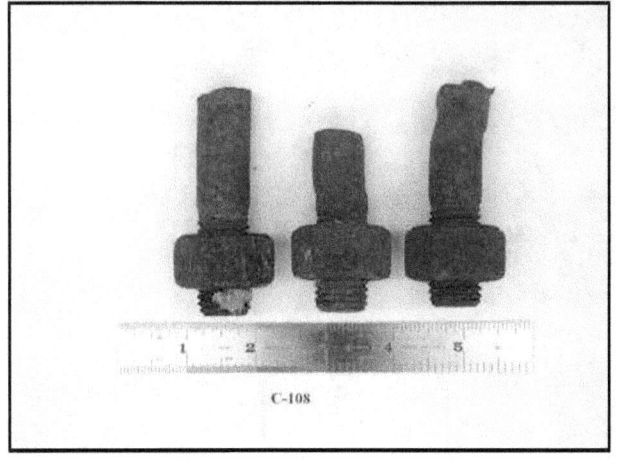
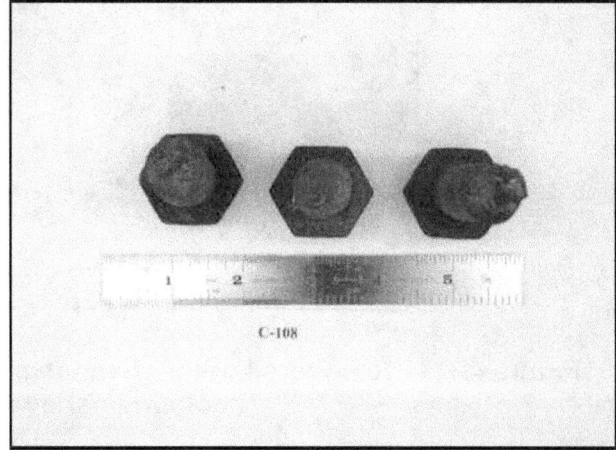
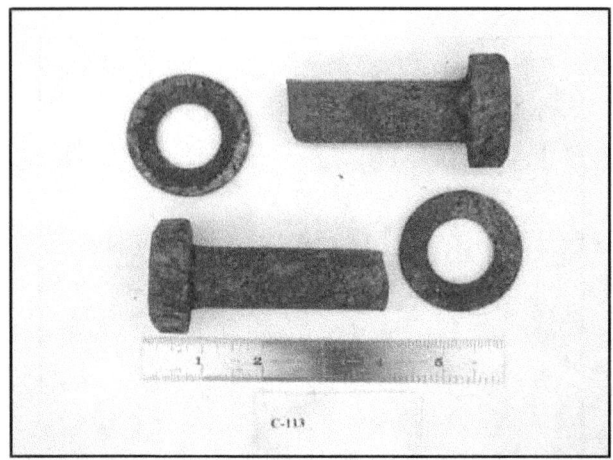
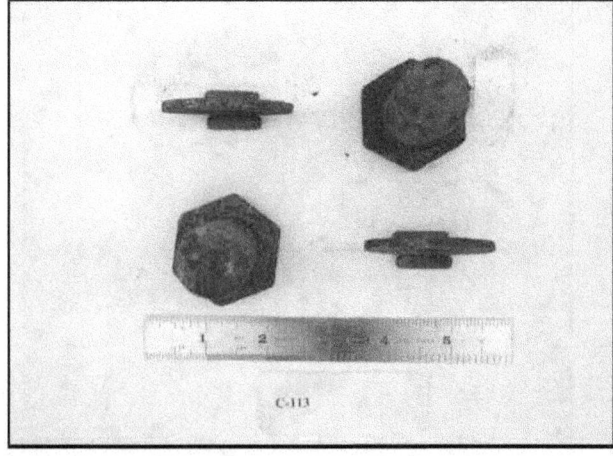

Source: NIST.

Figure A–13. Examples of recovered bolts from various samples.

Square tubular piece
Sample C-103

Rectangular tubular piece
Sample C-135

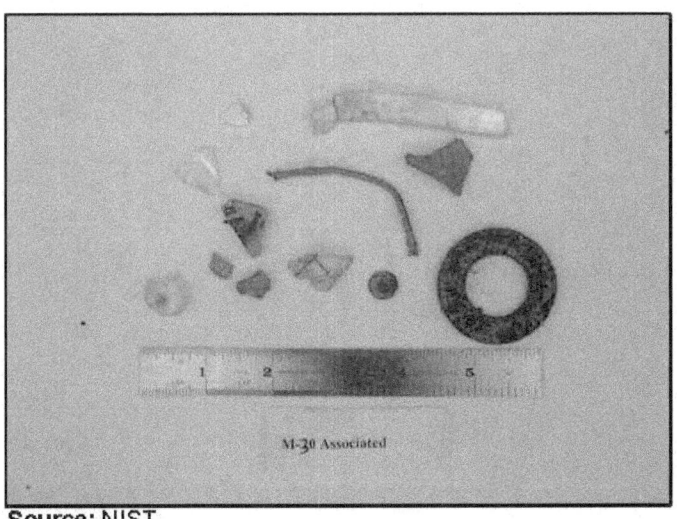
Assorted pieces from within column
Sample M-30 Associated

Source: NIST.

Figure A–14. Examples of miscellaneous materials recovered.

www.ingramcontent.com/pod-product-compliance
Lightning Source LLC
Chambersburg PA
CBHW080305180526
45167CB00006B/2679